智能微型运动装置（Micromouse）技术与应用系列丛书
工程实践创新项目（EPIP）教学模式系列教材

迷宫机器人仿真与设计

（中英双语版）

王　超　花玉香　宋立红◎主　编
　　　　　高　艺　郑勇峰◎副主编
刘　佳　严靖怡　杨迎娣◎编　译

中国铁道出版社有限公司
CHINA RAILWAY PUBLISHING HOUSE CO., LTD.

内 容 简 介

本书以天津启诚伟业科技有限公司提供的 TQD-OC V2.0 迷宫机器人虚拟仿真评测系统为载体，通过"渊源与发展"、"仿真基础"、"仿真设计"和"仿真实践"，讲述了虚拟仿真迷宫机器人的发展、仿真设计、开发环境、功能调试，以及迷宫机器人虚实交互、智能算法解析、应用拓展等内容。同时，本书附录提供了 2022 年首届世界职业院校技能大赛——迷宫机器人赛项规则、"迷宫机器人仿真与设计"国际实训课程标准等教学资源。

本书在重要的知识点、技术（技能）点和素养点上，配有丰富的视频、图片、文本等资源，学习者可以通过扫描书中二维码获取相关信息。

本书适合作为高等学校自动化类专业综合与创新实践教学的教材，也可作为职业院校电子与信息大类、自动化类专业开展课程教学、职业启蒙、科技活动和特色教育的教材，还可作为相关工程技术人员培训用书及迷宫机器人爱好者的参考书。

图书在版编目（CIP）数据

迷宫机器人仿真与设计：汉、英 / 王超，花玉香，宋立红主编 . — 北京：中国铁道出版社有限公司，2024.6
（智能微型运动装置（Micromouse）技术与应用系列丛书）
工程实践创新项目（EPIP）教学模式系列教材
ISBN 978-7-113-30807-0

I.①迷… II.①王…②花…③宋… III.①智能机器人 - 程序设计 - 教材 - 汉、英 IV.① TP242.6

中国国家版本馆 CIP 数据核字（2024）第 070034 号

书　　　名：	迷宫机器人仿真与设计（中英双语版）
作　　　者：	王　超　花玉香　宋立红

策　　　划：	何红艳	编辑部电话：	（010）63560043
责任编辑：	何红艳　绳　超		
封面设计：	刘　颖		
责任校对：	刘　畅		
责任印制：	樊启鹏		

出版发行：	中国铁道出版社有限公司（100054，北京市西城区右安门西街 8 号）
网　　址：	https://www.tdpress.com/51eds/
印　　刷：	北京盛通印刷股份有限公司
版　　次：	2024 年 6 月第 1 版　2024 年 6 月第 1 次印刷
开　　本：	787 mm×1 092 mm　1/16　印张：24　字数：409 千
书　　号：	ISBN 978-7-113-30807-0
定　　价：	79.80 元

版权所有　侵权必究

凡购买铁道版图书，如有印制质量问题，请与本社教材图书营销部联系调换。电话：（010）63550836
打击盗版举报电话：（010）63549461

作者简介

王 超

天津大学，教授，教育部高等学校自动化类专业教学指导委员会委员。主持包括国家自然科学基金重大仪器专项等国家及省部级项目10余项；获省部级科技奖励一等奖2项、二等奖3项。主持的"面向酿造过程的复杂系统控制虚拟仿真实验"2023年获批国家级一流课程。以第一主编出版教材7本，其中《智能鼠原理与制作（中英双语版）》系列教材入选首批"十四五"职业教育国家规划教材。获高等教育（本科）国家级教学成果奖二等奖1项、天津市教学成果奖一等奖1项、中国自动化学会高等教育教学成果奖二等奖1项。获宝钢优秀教师奖、教育部新世纪人才和天津市青年科技奖。主持完成的教育部产学合作协同育人项目入选2021年度教育部产学合作协同育人项目优秀项目案例。

花玉香

天津渤海职业技术学院党委副书记、院长。主持中国特色高水平高职院校和专业群建设；主持教育部虚拟仿真实训基地培育项目建设，并获批教育部高等学校科学研究发展中心项目——绿色生态化工虚拟仿真实训基地建设路径与成效研究；主持天津市职业教育化工安全技术专业教学资源库建设；主持完成天津市高技能人才培养基地；主持世界技能大赛6个天津市集训基地建设，主持申报开放性产教融合石油化工实践中心建设；注重教育教学成果凝练，构建了"1+3+4"思政育人体系，牵头实施"双栖制"教师团队建设，获得多项省部级教学成果奖。

宋立红

天津启诚伟业科技有限公司董事长，启诚迷宫机器人创始人，高级工程师，20余年专注致力于智能微型运动装置（迷宫机器人）的软硬件开发、设计、生产、服务工作。近三年作为启诚技术团队带头人主持国家及天津市科委科研项目3项。近三年发表论文3篇、主编专著及教材6册，其中2021年出版的《智能鼠原理与制作（中英双语版）》系列教材，入选首批"十四五"职业教育国家规划教材。从2016年至今，积极响应国家"一带一路"倡议和中国企业"走出去"，启诚迷宫机器人伴随着鲁班工坊走出国门与世界共享，目前泰国、印度、印尼、巴基斯坦、柬埔寨、尼日利亚、埃及、乌兹别克斯坦的鲁班工坊院校，均采纳《智能鼠原理与制作（中英双语版）》系列教材，开设相关国际教学课程，为培养"一带一路"当地国智能技术人才做出努力和贡献。

高 艺

南开大学电子信息工程学院硕士生导师,电子信息实验教学中心副主任。先后参与多项"国家高技术研究发展计划(863计划)项目""天津市科技支撑计划重点项目"以及横向科研项目。发表相关科研及教改论文16篇,其中多篇被EI检索。作为指导教师带学生组队参加全国大学生电子设计竞赛、天津市大学生IEEE电脑鼠竞赛等多项国家级、省部级学生竞赛,并屡获佳绩。

郑勇峰

天津渤海职业技术学院机电工程学院院长。获得天津市教学成果奖一等奖,中国石油和化工教育教学成果一等奖、二等奖。编写完成机电一体化国际化专业标准并获得泰国职业教育委员会认证。发表论文10余篇,获得专利30余项,主编教材5部。

刘 佳

南开大学外国语学院博士。国家级精品课"大学英语"和"科研方法论"主讲教师。曾荣获南开大学教学基本功大赛一等奖,天津市教学基本功大赛二等奖。参编口语教材、翻译资格考试教材、写作教材和词典等数部,翻译原版英文书籍数部。

严靖怡

美国西北大学传播学硕士,现任天津启诚伟业科技有限公司总经理特别助理。参加柬埔寨鲁班工坊Micromouse项目翻译,在柬期间,辅导当地学生学习中国迷宫机器人技术,获得柬埔寨国立理工学院荣誉证书。2021年编译《智能鼠原理与制作(中英双语版)》系列教材入选首批"十四五"职业教育国家规划教材。

杨迎娣

天津渤海职业技术学院国际交流与合作办公室教师。2022年指导学生获中国教育电视台外研社杯英语演讲大赛三等奖2项,主持局级课题3项,获资金资助1项,结题成果获天津市二等奖1项。

前言

"迷宫机器人"又称"智能鼠",英文为 Micromouse,其技术融合了智能控制算法、多传感器融合、3D虚拟仿真、智能图像识别、物联网与通信等相关技术,全面对标现代智能科技产业发展。

本书为"智能微型运动装置(Micromouse)技术与应用系列丛书"之一,也是"工程实践创新项目(EPIP)教学模式系列教材"之一。其中的《智能鼠原理与制作(中英双语版)》系列教材于2023年入选首批"十四五"职业教育国家规划教材,输出到"一带一路"鲁班工坊,为培养当地智能化技术人才做出努力和贡献。

近年来,全球数字化进程加速演进,新一代信息技术与数字技术的广泛应用和深度融合,正逐步"重塑"教育的理念、模式与形态。把人工智能、虚拟仿真等数字技术越来越多地应用到教学和竞赛中,使学生知识获取的效率、质量、精准度得到进一步提升。天津大学、南开大学、天津渤海职业技术学院、天津启诚伟业科技有限公司,产教融合、校企合作,以工程实践创新项目(EPIP)教学模式,对迷宫机器人竞赛进行创新改革,构建线上线下相结合、虚拟仿真与现实互动的竞赛模式,从竞赛的内容及方式上激发学生创新活力,提升教师信息化教学融合应用水平,以数字化转型赋能职业教育"中、高、本"贯通高质量发展,对于后期迷宫机器人竞赛的开展和走进课堂、融入教学起到关键性的作用。为了将虚拟仿真迷宫机器人的成果进一步推广应用,特组织编写了适用于高等教育和职业教育的《迷宫机器人仿真与设计(中英双语版)》教材。本书从虚拟仿真迷宫机器人的"渊源与发展"、"仿真基础"、"仿真设计"和"仿真实践"讲述了迷宫机器人的发展、设计、开发、功能调试,以及迷宫机器人虚实交互、智能算法解析、应用拓展等内容,以丰富学生的工程实践创新能力,拓宽专业视野,内化形成良好的职业素养,助力教学数字化转型发展。

本书具有以下创新和特色:

1. 校企合作双向赋能,充分体现产教融合互融共生的合作思想

本书按照学生职业能力成长的过程进行设计,建立校企合作的教材开发团队,由学校教师与企业技术人员共同制定课程标准、共同编写。本书所选案例均来自真实的工程项目,编者来自国内长期从事迷宫机器人设计与开发、国际迷宫机器人竞赛获奖的院校和企业。特别是2022年8月在首届世界职

业院校技能大赛上，天津渤海职业技术学院入选"迷宫机器人"赛项承办院校，天津启诚伟业科技有限公司入选"迷宫机器人"赛项合作企业。校企双方充分发挥产教融合、校企合作精神，共同研发设计出TQD虚拟仿真迷宫机器人竞赛评测系统，在竞赛中得到广泛应用和专家认可。该系统遵循国际竞赛标准，结合教育教学中实际情况创新改革，形成具有中国特色的虚拟仿真迷宫机器人竞赛标准。

2. 抓住"一带一路"共建国家鲁班工坊新机遇，助力职教教学资源国际化建设

"推动共建'一带一路'高质量发展"是党的二十大报告中提出的重要举措。打造中国职业教育国际化品牌，是推进现代职业教育体系改革的重要任务。截止到2023年，迷宫机器人系列双语教材，服务于泰国、印度、印尼、巴基斯坦、柬埔寨、尼日利亚、埃及、乌兹别克斯坦8个国家9所"鲁班工坊"的教学课程建设。本书的国际课程标准由天津渤海职业技术学院与天津启诚伟业科技有限公司牵头开发，迷宫机器人软硬件教学装备伴随鲁班工坊走出国门与世界分享，产教融合效果显著，为"一带一路"共建国家提供丰富实践教学资源，服务当地技术技能人才培养。

本书由天津大学教授王超，天津渤海职业技术学院党委副书记、院长花玉香，天津启诚伟业科技有限公司总经理、启诚迷宫机器人创始人宋立红担任主编，南开大学副教授高艺、天津渤海职业技术学院教授郑勇峰担任副主编。南开大学外国语学院刘佳、天津启诚伟业科技有限公司总经理特别助理严靖怡、天津渤海职业技术学院杨迎娣担任本书的编译。

本书在编写过程中，参阅了大量书籍，得到了天津大学、天津渤海职业技术学院、南开大学等相关院校教授专家的大力支持，同时，天津启诚伟业科技有限公司提供了企业实际工程案例、思维导图、二维码视频及动画课程资源，中国铁道出版社有限公司支持出版，在此一并表示衷心感谢。

尽管我们在探索教材特色的建设方面有了一定的突破，但限于编者的经验、水平以及时间，书中难免存在不妥和疏漏之处，恳请各位同仁、读者不吝赐教。

编　者
2024年1月

第〇篇　迷宫机器人的渊源与发展 ... 001
项目一　迷宫机器人的发展渊源 ... 003
项目二　迷宫机器人与世校赛的渊源 ... 010

第一篇　迷宫机器人仿真基础 ... 013
项目一　Python开发入门 ... 015
任务一　认识Python开发环境 ... 015
任务二　掌握Python数据类型 ... 025
任务三　学习Python函数和类 ... 037
任务四　掌握Python扩展库 ... 042

项目二　学习环境搭建 ... 047
任务一　了解Ubuntu系统 ... 047
任务二　学习ROS2的安装 ... 054
任务三　学习Gazebo安装与集成 ... 058

项目三　ROS2入门 ... 062
任务一　了解ROS2工作空间与常用命令 ... 062
任务二　学习ROS2节点 ... 066
任务三　学习ROS2话题 ... 070

项目四　Gazebo入门 ... 077
任务一　掌握Gazebo基本操作 ... 077
任务二　学习URDF模型搭建 ... 082

第二篇　迷宫机器人仿真设计 ... 089
项目一　学习迷宫机器人制作 ... 091
任务一　制作虚拟仿真迷宫机器人 ... 091
任务二　制作虚拟仿真迷宫 ... 095
任务三　学习虚拟仿真系统启动 ... 101

项目二　了解迷宫机器人基础功能 105
　　任务一　了解两轮差速驱动 105
　　任务二　学习激光雷达检测 114
　　任务三　掌握迷宫机器人循迹运行 118
项目三　掌握迷宫机器人智能控制 122
　　任务一　学习坐标计算 122
　　任务二　学习路口检测 127
　　任务三　掌握转弯控制 129

第三篇　迷宫机器人仿真实践 137

项目一　学习迷宫机器人智能算法 139
　　任务一　了解迷宫搜索策略 139
　　任务二　掌握迷宫信息的存储与读取 144
　　任务三　掌握迷宫最优路径规划 147
项目二　迷宫机器人竞赛虚拟仿真实训 151
　　任务一　学习虚拟仿真评测系统 151
　　任务二　了解虚拟仿真竞赛 156

附录 159

附录A　2022年首届世界职业院校技能大赛——迷宫机器人赛项规则 159
附录B　"迷宫机器人仿真与设计"国际实训课程标准 164
附录C　教学内容与课时安排 167
附录D　专业词汇中英对照表 168

第〇篇　迷宫机器人的渊源与发展

迷宫机器人又称智能鼠（Micromouse），英文名为maze robot。迷宫机器人技术融合了智能控制算法、多传感器融合、3D虚拟仿真、智能图像识别、物联网与通信等相关技术，全面对标现代智能科技产业发展，迷宫机器人竞赛是展现产学研合作成果，推动赛事成果转化的重要载体。

迷宫机器人竞赛在国际上已经有50多年的历史，分为古典现实竞赛和虚拟仿真竞赛。迷宫机器人竞赛构建了线上线下相结合、虚拟仿真与现实互动的混合竞赛模式，从迷宫机器人竞赛的内容与方式方法上创新竞赛新形态，激发大学生新活力，以数字化转型赋能高等教育和职业教育高质量发展。

古典现实竞赛主要内容是迷宫机器人从起点出发，在不受人为操纵影响的条件下在未知的迷宫中，自主搜索迷宫找到终点，并选择出最短的一条路径进行冲刺。竞赛成绩根据搜索迷宫的时间和冲刺时间给出，竞赛迷宫遵照电气电子工程师学会（IEEE）的国际标准。

当前，随着全球数字化进程加速推进，新一代信息技术与数字技术的广泛应用和深度融合，逐步"重塑"教育的理念、模式与形态。天津启诚伟业科技有限公司与天津渤海职业技术学院进行校企合作，遵循国际竞赛标准，结合教育教学中实际情况进行创新改革，共同研发设计出TQD虚拟仿真迷宫机器人竞赛评测系统，形成具有中国特色的虚拟仿真迷宫机器人竞赛标准。

虚拟仿真竞赛主要内容是参赛者首先在计算机系统中搭建3D迷宫以及3D机器人环境、部署虚拟仿真竞赛环境，然后3D机器人在未知的复杂多变迷宫中，进行遍历搜索，通过智能控制算法计算与评估出最优路径，以最快的速度从起点冲刺到终点的比赛过程。

虚拟仿真迷宫机器人竞赛面向电子信息技术和装备制造等多个产业发展需求，以智能控制应用为主题，结合信息技术和智能控制技术国际标准、职业标准和岗位要求，对智能控制项目实施过程中所涉及的知识和技能进行考核。竞赛融合了嵌入式微控制器技术及应用、传感器技术及应用、智能控制算法应用、3D虚拟仿真技术应用、人工智能与计算技术应用等关键技术、核心知识和核心技能成果。赛项充分展示了人工智能与机器人学科发展的新技术成果，引领高素质技术技能型人才培养方向和职业院校专业升级，提升专业建设的能力，推动赛事成果转化和产学研国际合作，紧贴市场需求，聚焦新技术新业态，搭建国际职业技术教育交流的高水平平台，助力国际职业技术院校高质量发展，传递中国职业教育办学理念，输出中国职业教育办学办赛经验，加强国际合作，实现共同提高。

本书将分别从3D建模、软件开发环境和编程等方面系统学习仿真迷宫机器人技术，并对仿真迷宫机器人的基本原理和实际操作方法进行具体说明。

项目一

迷宫机器人的发展渊源

学习目标

1. 了解迷宫机器人的起源。
2. 了解迷宫机器人在国际和国内竞赛的发展历程。

1938年,美国密歇根州的数学家香农(Shannon)完成了《继电器和开关电路的符号分析》的论文。由于布尔代数只有0和1,恰好与二进制对应,香农将它运用于以脉冲方式处理信息的继电器开关,从理论到技术彻底改变了数字电路的设计方向。因此,这篇论文在现代数字计算机史上具有划时代的意义。

1948年,香农又发表了一篇至今还在闪烁光芒的论文《通信的数学理论》,从而给自己赢得了"信息论之父"的桂冠。

1956年,他参与发起了达特茅斯人工智能会议,成为这一新学科的开山鼻祖之一。他不仅率先把人工智能运用于计算机下棋方面,而且还发明了一个能自动穿越迷宫的"迷宫机器人",以此证明计算机可以通过学习提高智能。

迷宫机器人的国际发展渊源如图0-1-1所示。

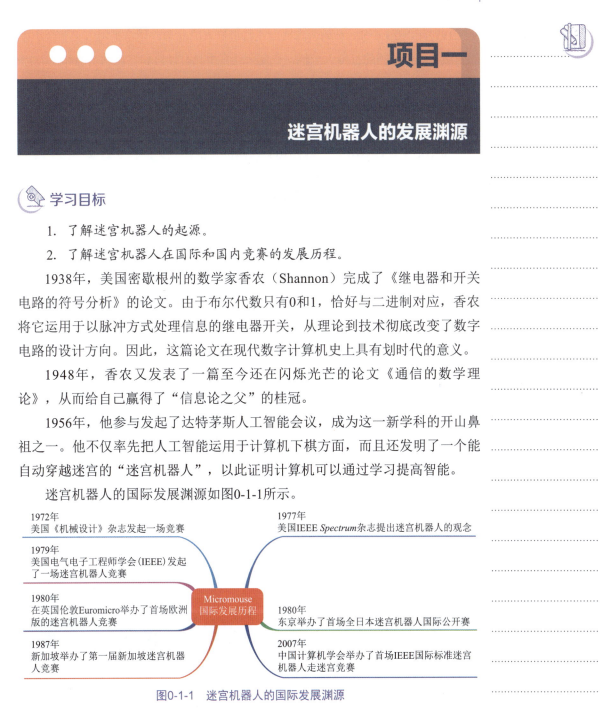

图0-1-1 迷宫机器人的国际发展渊源

1972年,《机械设计》杂志发起了一场竞赛。在竞赛中,仅由捕鼠器弹簧驱动的机械鼠,不停地与其他参赛鼠竞赛,以判断哪个机械鼠能够沿着跑道跑出最长的距离。

1977年，IEEE *Spectrum*杂志提出迷宫机器人的观念。迷宫机器人是一个小型的由微处理器控制的机器人车辆，在复杂迷宫中具有译码和导航的功能和能力。

1979年，电气电子工程师学会（IEEE）通过其*Spectrum*杂志发起了一场迷宫机器人竞赛，奖励能够在最短时间内自主走出迷宫的迷宫机器人的设计者1 000美元。

1980年，首场世界迷宫机器人大赛在日本举行，之后，又有多个比赛被创办，如1980年英国迷宫机器人大赛，1987年新加坡举办的IES迷宫机器人大赛和2007年中国计算机学会举办的大学生迷宫机器人走迷宫竞赛等。

从最初1972年的机械电子鼠发展到现在的迷宫机器人，如图0-1-1所示，迷宫机器人竞赛经过了50多年沧海桑田的蜕变。参加竞赛的选手从开始仅限于哈佛大学、麻省理工学院等世界知名学府的研究生，发展到从研究型大学到应用技术大学再到职业院校的学生甚至是中小学生。多教育层面都采纳了迷宫机器人为教学载体，培养学生们的工程素养以及科技创新意识、动手设计能力。各类迷宫机器人竞赛也如雨后春笋般蓬勃发展。目前迷宫机器人竞赛已经成为应用于不同教育阶段的国际创新型学生竞赛。

从2007年至今，迷宫机器人在中国经历了十余年的成长发展历程，如图0-1-2所示。2007年，天津启诚伟业科技有限公司将这项国际赛事引进天津，以中国先进的教育模式"工程实践创新项目"为核心理念，对迷宫机器人走迷宫竞赛进行本土化创新改革，助力了迷宫机器人竞赛在中国的蓬勃开展，对于迷宫机器人技术走进课堂融入教学起到关键性的引领作用。

图0-1-2 迷宫机器人在中国的发展

竞赛对于满足产业优化升级，开阔国际视野，掌握实践与创新经验培养高技术、高技能人才，起到了引领推动作用。迷宫机器人在中国从本科院校竞赛到职业院校大赛，再到普职融通国际挑战赛，积累了丰富的竞赛经验和优秀的技术积淀，如图0-1-3所示。

图0-1-3　竞赛纪实照片

十余年来，中国的迷宫机器人竞赛不断创新国际发展新思路，从最初的"简单模仿"学习，发展到目前的"互学互鉴"，逐步搭建起国际交流合作的新平台，先后经历了学习借鉴、蜕变升华和引领辐射三个阶段。

学习借鉴：2015年天津大学生代表队征战美国第30届APEC世界Micromouse竞赛，如图0-1-4所示，取得了世界排名第六的好成绩。2017年至2018年，天津启诚伟业科技有限公司全额资助了在天津大学生迷宫机器人竞赛上获得企业命题赛冠军队，到日本东京参加第38届和第39届全日本迷宫机器人国际公开赛，如图0-1-5所示，学习借鉴国际迷宫机器人先进技术，结识众多迷宫机器人业界专家教授，有助于提升中国迷宫机器人技术的发展。

图0-1-4　中国天津代表队远征美国参加国际大赛

图0-1-5　中国天津代表队远赴日本参加国际大赛

蜕变升华：迷宫机器人大赛在中国进行创新改革，设计了一系列从易到难的启诚迷宫机器人教学平台，满足"中、高、本、硕"不同学习阶段学生学习应用。从2016年开始，先后邀请美国麻省理工学院的戴维·欧腾教授、中国台湾龙华科技大学苏景晖教授、新加坡义安理工学院黄明吉教授、英国伯明翰城市大学彼得·哈里森教授、日本迷宫机器人国际公开赛组委会秘书长中川友纪子先生等迷宫机器人专家和来自泰国、印度、印度尼西亚、巴基斯坦、柬埔寨等国际"鲁班工坊"师生，以及来自中国天津、北京、河南、河北等国内省市精英级代表队，先后加盟中国IEEE电脑鼠走迷宫国际邀请赛，如图0-1-6至图0-1-8所示。国际选手通过参加中国比赛，对中国竞赛标准、竞赛规则、竞赛模式和竞赛理念有了更深层次的了解和认同，从而切实推动国际化的交流与合作，达到互学互鉴的目的。

图0-1-6　第三届IEEE电脑鼠走迷宫国际邀请赛

图0-1-7 "启诚杯"第四届IEEE电脑鼠走迷宫国际邀请赛

图0-1-8 第五届"启诚杯"智能鼠走迷宫国际邀请赛

引领辐射：教育对外开放是我国改革开放事业的重要组成部分，随着"一带一路"倡议的推进，2016年以来在教育部指导下，先后启动了海外鲁班工坊国际项目，迷宫机器人作为中国优秀的教育装备，伴随着鲁班工坊走出国门与世界分享。从2016年至今，启诚迷宫机器人来到泰国、印度、印尼、巴基斯坦、柬埔寨、尼日利亚、埃及、乌兹别克斯坦等国家，开展迷宫机器人竞赛的推广和课程培训，如图0-1-9至图0-1-14所示，受到了共建国家师生的一致青睐。迷宫机器人成为连接世界的纽带与桥梁。

图0-1-9　印度鲁班工坊开展迷宫机器人培训课程

图0-1-10　2016年泰国鲁班工坊开展迷宫机器人培训课程

图0-1-11　2017年印度尼西亚鲁班工坊开展迷宫机器人培训课程

图0-1-12　2018年巴基斯坦鲁班工坊开展迷宫机器人培训课程

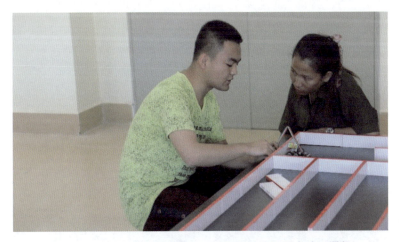

图0-1-13　2018年柬埔寨鲁班工坊开展迷宫机器人培训课程

图0-1-14　2020年埃及鲁班工坊开展迷宫机器人培训课程

项目二

迷宫机器人与世校赛的渊源

 学习目标

了解迷宫机器人与世校赛的渊源。

2022年教育部决定在天津举办世界职业技术教育发展大会,此次大会为中国政府首次主办的国际性职业教育会议。同期举办世界职业院校技能大赛,以"增进交流、深化合作、创新发展"为原则,每两年举办一次大会,逐步将大会打造成职业教育领域具有重要国际影响力的机制性会议,推动全球职业教育互学互鉴、共商共享,助力构建人类命运共同体。

2022年8月世界职业院校技能大赛(以下简称"世校赛")执行委员会公布《关于公布首届世界职业院校技能大赛赛项承办校和合作企业的通知》(赛执委函〔2022〕38号),天津渤海职业技术学院入选首届世校赛"迷宫机器人"赛项承办院校。天津启诚伟业科技有限公司入选首届世校赛"迷宫机器人"赛项合作企业,如图0-2-1所示。

图0-2-1 首届世校赛"迷宫机器人"赛项

截止到2022年,已有8个国家9所鲁班工坊院校,开设了迷宫机器人对应专业课程并建设了EPIP实训室。包括:泰国鲁班工坊机电一体化技术专业;印度鲁班工坊工业机器人专业；印度尼西亚鲁班工坊电子技术应用专业；巴基斯坦鲁班工坊自动化控制专业；柬埔寨鲁班工坊机电一体化技术专业；尼日利亚鲁班工坊电气电子工程专业；两所埃及的鲁班工坊对应专业分别为工业机器人和自动化控制；乌兹别克斯坦鲁班工坊对应开设信息技术专业课程。迷宫机器人具有多专业组群建设应用、多产业场景聚焦应用等特点，结合信息技术和智能控制技术国际标准、职业标准和岗位要求，对智能控制项目实施过程中所涉及的知识和技能进行考核。

2022年8月11日，由教育部、发展改革委员会、科学技术部、工业与信息化部、天津市人民政府等35家单位主办，天津市教育委员会、天津市教育科学研究院、天津渤海化工集团有限公司承办，天津渤海职业技术学院、天津启诚伟业科技有限公司协办的首届世界职业院校技能大赛"迷宫机器人"赛项鸣锣开赛。来自中国、赤道几内亚、贝宁、也门、埃塞俄比亚、苏丹、刚果、塞拉利昂、津巴布韦等国家的中外选手组成"混合战队"，在线上展开激烈角逐。

来自西非贝宁共和国的参赛选手卡普诗诗·赛格农·约翰（见图0-2-2）和天津渤海职业技术学院学生于欣令组成的中外混合赛队，最终获得迷宫机器人竞赛金牌。

图0-2-2 "迷宫机器人"赛项参赛选手

约翰在接受媒体采访的时候透露，自己从小就对科技抱有好奇心，今年是他在中国留学的第三年，在天津学习期间，他觉得自己很幸运，接触到的都是好老师。在与于欣令同学一起合作比赛的过程，为观赛者们上演了一场微缩版的"速度与激情"。通过这次比赛不仅提高了技能，而且通过彼此的切磋和交流增进了友谊。他坚信中国的职业技术教育会越来越好，会给这个世界带来更多的价值。

赛项裁判南开大学副教授高艺说，近年来3D虚拟仿真技术备受瞩目，2022年的虚拟仿真竞赛环节受到国际师生的青睐。只要有网络，选手都可以通过TQD-OC V2.0虚拟仿真评测系统，不受地域、时空限制，实时进行公平、公正、公开同场竞技。迷宫机器人正从智能运动装置逐步走向数字孪生技术，以服务产业和技术发展为导向，主动与国际职业教育接轨，发挥鲁班工坊桥梁纽带作用，服务"一带一路"建设，立足市场需求深化产学研合作，进一步打造全国职教高地。

思考与总结

1. 在迷宫机器人竞赛中包含哪两个赛项？它们之间区别是什么？
2. 迷宫机器人虚拟仿真评测系统作用是什么？
3. 迷宫机器人竞赛已经成为应用于不同教育阶段的国际创新型学生竞赛。迷宫机器人作为教学载体，培养学生们工程素养以及科技创新意识动手设计能力。

第一篇 迷宫机器人仿真基础

　　迷宫机器人仿真是使用软件和程序仿真迷宫机器人的整个运行过程。本篇主要讲解在仿真时使用的编程语言以及软件运行环境。

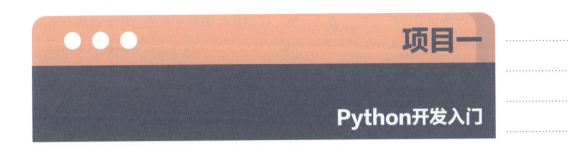

项目一 Python开发入门

> **学习目标**
>
> 1. 了解Python与VSCode的安装与使用方式。
> 2. 了解Python的基本语法。
> 3. 了解Python常用的标准库与第三方库。

任务一 认识Python开发环境

本任务是学习搭建Python开发环境，并掌握基础语法。

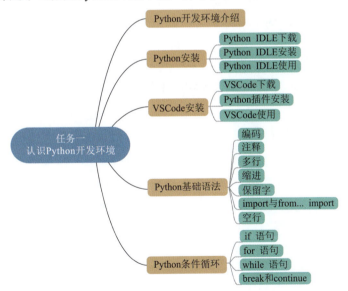

一、Python开发环境介绍

Python是一种跨平台的计算机程序设计语言，是面向对象、解释型、动态数据类型的高级语言，可以用于Windows、Linux和Mac平台上，图标如图1-1-1所示。

图1-1-1 Python图标

Python的语法简单，使用方便，即便是非软件专业的初学者也很容易上手，越来越多的人开始使用Python进行软件开发。相对于其他编程语言来说，Python有以下几个优点：

① 开源免费。使用Python进行开发或者发布程序不需要支付任何费用，也不用担心版权问题，即使作为商业用途，Python也是免费的。

② 语法简单。相比较C/C++、C#、VB、Java等语言，Python对代码格式的要求较低，编写程序时不需要在细枝末节上花费太多精力。

③ 语言高级。Python是对C语言的封装，屏蔽了很多底层细节，比如自动管理内存，需要时自动分配，不需要时自动释放。

④ 移植性好。可应用于多个平台上。

⑤ 面向对象的编程语言。具有面向对象的继承、多态、封装等特性。

⑥ 扩展性强。有广泛的扩展库，可以帮助完成各种各样的程序。覆盖了文件I/O、数值计算、GUI编程、网络编程、数据库访问等绝大部分应用场景。

Python的发展历史经历了Python 2和Python 3，由于差异较大，并且官方已经在2020年停止对Python 2的维护，因此推荐使用Python 3进行开发。本书后续内容中所讲的Python均指Python 3。

二、Python安装

Python开源免费，用户可以到Python官网直接下载并安装。

1. Python IDLE下载

目前Python的最新正式版是3.11.x，在学习Python时推荐使用最新的正式版，但在开发项目时，推荐降低一个版本，选择3.10.x，这是因为一些第三方库可能还未适配最新正式版，容易出现各种各样的错误。

根据自己的计算机系统类型选择适合的安装程序，如图1-1-2所示。

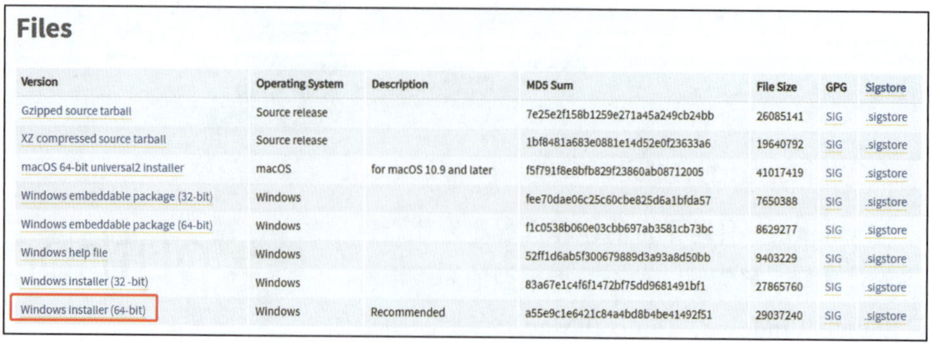
Files

Version	Operating System	Description	MD5 Sum	File Size	GPG	Sigstore
Gzipped source tarball	Source release		7e25e2f158b1259e271a45a249cb24bb	26085141	SIG	.sigstore
XZ compressed source tarball	Source release		1bf8481a683e0881e14d52e0f23633a6	19640792	SIG	.sigstore
macOS 64-bit universal2 installer	macOS	for macOS 10.9 and later	f5f791f8e8bfb829f23860ab08712005	41017419	SIG	.sigstore
Windows embeddable package (32-bit)	Windows		fee70dae06c25c60cbe825d6a1bfda57	7650388	SIG	.sigstore
Windows embeddable package (64-bit)	Windows		f1c0538b060e03cbb697ab3581cb73bc	8629277	SIG	.sigstore
Windows help file	Windows		52ff1d6ab5f300679889d3a93a8d50bb	9403229	SIG	.sigstore
Windows installer (32 -bit)	Windows		83a67e1c4f6f1472bf75dd9681491bf1	27865760	SIG	.sigstore
Windows installer (64-bit)	Windows	Recommended	a55e9c1e6421c84a4bd8b4be41492f51	29037240	SIG	.sigstore

图1-1-2　Python下载

下载地址请自行在搜索引擎中搜索获得。

需要说明的是，在Ubuntu 22.04中，已经集成了3.10.6，不需要额外安装了。

2. Python IDLE安装

Python的安装文件非常小，仅27 MB左右，作为演示，这里选择"Windows installer (64-bit)"在Windows中进行安装。

Python的安装过程和普通软件类似，如图1-1-3所示，推荐选中两个复选框。

图1-1-3　检查管理员权限及添加路径

推荐移除路径的长度限制，如图1-1-4所示，否则可能会因为某些程序文件的路径过长，导致失败。

3. Python IDLE使用

Python不会创建桌面快捷方式，安装完毕后，可以在开始菜单中找到Python，如图1-1-5所示。单击IDLE启动Python，如图1-1-6所示。

图1-1-4　移除路径长度限制

图1-1-5　查找Python

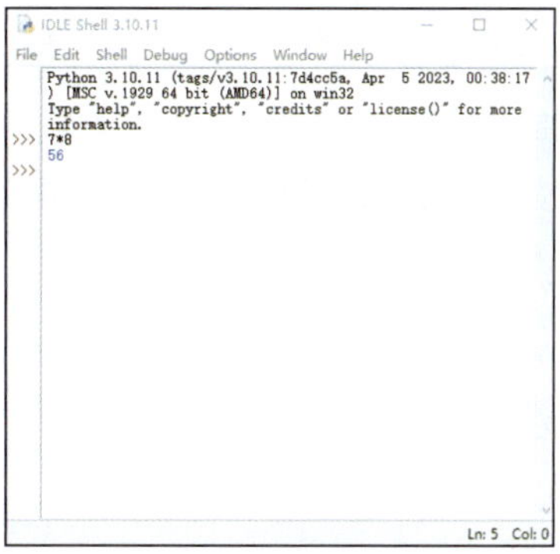

图1-1-6　使用Python IDLE

在IDLE中键入程序，按【Enter】键执行。

三、VSCode安装

如果只是编写简单的程序，在Python自带的开发环境中写代码是可以的。但对于专业的程序员来说，其编写的程序比较复杂，在Python自带的开发环境中编写代码就有些捉襟见肘了，尤其是编写面向对象的程序，无论是代码提示功能还是出错信息的提示功能远没有专业开发环境的功能强大。

VSCode是一款微软出品的轻量级编辑器，它本身只是一款文本编辑器，所有的功能都是以插件扩展的形式存在的，需要什么功能就安装对应的扩展，包括Python、HTML、C/C++、数据库等，非常方便。VSCode支持各大主流操作系统，包括Windows、Linux和MacOS。所以，本书选择VSCode作为主要的编辑器来使用。

1. VSCode下载

VSCode完全免费，访问官方网站自行下载安装，根据自己的计算机系统类型选择适合的安装程序，如图1-1-7所示。

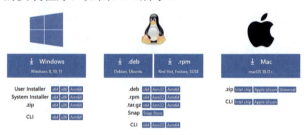

图1-1-7　VSCode下载

对于Windows用户：

如果是普通用户，选择"User Installer"中的x64、x86或Arm64；

如果是管理员用户，选择"System Installer"中的x64、x86或Arm64。

2. Python插件安装

对于Ubuntu用户：还可以在系统的软件商店中，搜索"Code"进行下载安装。

VSCode的安装过程和普通软件相同，比较简单，不再赘述。

进行Python开发就需要安装Python扩展，如图1-1-8所示，在插件中搜索"python"。

安装完成后就具备了Python开发能力，比如调试、语法高亮、代码跳转、智能提示、自动完成、单元测试、版本控制等，极大地提高了Python的开发效率。

图1-1-8　安装Python扩展

辅助Python开发的插件还有很多,例如indent-rainbow、Jupyter、Image preview等。

3. VSCode使用

VSCode会自动识别计算机中的Python解释器。创建一个脚本文件demo.py,键入程序,选择在终端中运行Python文件,如图1-1-9所示。

图1-1-9　运行VSCode

Python和VSCode安装完毕后,就可以开始进行Python的基础编程了。

四、Python基础语法

同其他编程语言一样,编写Python程序需要遵守Python语法规则,Python语言与C和Java等语言有许多相似之处。但是,也存在一些差异。下面介绍一些基础语法。

1. 编码

默认情况下,Python源码文件以UTF-8编码,所有字符串都是unicode字符串。当然也可以为源码文件指定不同的编码。

```
# -*- coding: cp-1252 -*-
```

2. 注释

注释是对代码的解释和说明,目的是让人们能够更加轻松地了解代码。

Python中单行注释以#开头，多行注释采用三对单引号（'''）或者三对双引号（"""）将注释括起来。Python解释器会自动跳过这部分内容，不予执行。

```
if 1+3>2:    # 这是单行注释
    """
    这是多行注释
    这是多行注释
    """
    '''
    也可以用三个单引号
    来进行多行注释
    '''
    print(1)
else:
print(0)
```

3. 多行

Python是解释型语言，程序通常是按照行来执行的，以新行作为上一行语句的结束符。若代码较长，或者为了美观，可以将一行语句拆分为多行，但要使用反斜杠"\"表示这是一条语句，反斜杠"\"在这里被称为连接符。

```
>>> a = 8+9+\
        10+12+\
        15
>>>a
    54
```

若多行是包含在"[]"，"()"，"{}"中的，就不需要使用连接符了。

```
>>> a = (8+9+
        10+12+
        15)
>>>a
    54
```

4. 缩进

Python最具特色的就是使用缩进来表示代码块，其他编程语言多使用花括号"{}"。缩进的空格数是可变的，但是同一个代码块的语句必须包含相同的缩进空格数。

```
if True:
    print(True)
else:
    print(False)
```

Python官方推荐每级缩进使用4个空格。

```
while 1:
```

```
    if 1:                   # 1级缩进，4个空格
        if 1:               # 2级缩进，8个空格
            print(1)        # 3级缩进，12个空格
        else:
            print(0)
    else:
        print(0)
```

5. 保留字

保留字即关键字。不能把它们用作任何标识符名称。Python的标准库提供了一个关键词模块，可以使用它来查看当前版本的所有保留字。

```
>>> import keyword
>>> keyword.kwlist
```

终端中返回如下信息：

```
['False', 'None', 'True', 'and', 'as', 'assert', 'async', 'await',
'break', 'class', 'continue', 'def', 'del', 'elif', 'else', 'except',
'finally', 'for', 'from', 'global', 'if', 'import', 'in', 'is', 'lambda',
'nonlocal', 'not', 'or', 'pass', 'raise', 'return', 'try', 'while',
'with', 'yield']
```

Python对大小写敏感，例如FALSE，就不是保留字。

```
>>> False
    False
>>> FALSE
Traceback (most recent call last):
  File "<pyshell#33>", line 1, in <module>
    FALSE
NameError: name 'FALSE' is not defined. Did you mean: 'False'?
>>> FALSE=1
>>> FALSE
    1
```

6. import与from…import

在Python中使用import或者from…import来导入扩展库。

将整个模块导入，格式为：import somemodule。

从某个模块中导入某个函数，格式为：from somemodule import somefunction。

从某个模块中导入多个函数，格式为：from somemodule import firstfunc, secondfunc, thirdfunc。

将某个模块中的全部函数导入，格式为：from somemodule import *。

还可以使用as表示别名，格式为：from somemodule import somefunction as newname。

```
>>> import os
>>> os.getcwd()
'D:\\Python3'
>>> from sys import getdefaultencoding
>>> getdefaultencoding()
'utf-8'
>>> import numpy as np
>>> np.ones((3,3))
array([[1., 1., 1.],
       [1., 1., 1.],
       [1., 1., 1.]])
```

7. 空行

空行并不是Python语法的一部分。书写时不插入空行，Python解释器运行也不会出错。但是空行的作用在于分隔两段不同功能或含义的代码，便于日后代码的维护或重构。

用空行分隔，表示一段新的代码的开始。没有从属关系的类和函数之间推荐用两行空行分隔；类的方法之间推荐用一行空行分隔，以突出显示。

```
def func1():
    pass


def func2():
    pass


class Demo:

    def func1():
        pass

    def func2():
        pass
```

五、Python条件循环

Python条件语句是通过一条或多条语句的执行结果（True或者False）来决定执行的代码块。

Python循环语句是通过条件语句的执行结果（True或者False）来决定是否循环执行代码块。

1. if 语句

if语句是一个条件语句，用于组成条件控制，用法：

```
if condition1:
```

```
    statement_block_1
elif condition2:
    statement_block_2
else:
    statement_block_3
```

if语句永远只会执行一个statement_block，其中elif和else部分均可以省略，也可以有多个elif，但else最多只能有一个。

if语句的执行顺序从上至下，判断condition的结果，只要某个condition为True，就执行内部的statement_block，其他的condition将不再判断。

```
a=8
if a>10:
    print('a>10')
elif 5<=a<=10:
    print('5<=a<=10')
else:
    print('a<5')
```

终端中输出如下信息：

```
5<=a<=10
```

if中常用的操作符见表1-1-1。

表1-1-1　if中常用的操作符

操 作 符	描　　述	操 作 符	描　　述
<	小于	>=	大于或等于
<=	小于或等于	==	等于，比较两个值是否相等
>	大于	!=	不等于

2. for语句

for语句是一个循环语句，循环可以遍历任何可迭代对象，如一个列表或者一个字符串，通常结合in使用，用法：

```
for variable in iterable:
    statements
else:
    statements
```

for语句可以单独使用，也可以结合else子句，与if语句不同，只有当for遍历成功后，才会执行else中的语句。

```
sites = ["Baidu", "Google","Jingdong","Taobao"]
for site in sites:
    print(site)
```

```
        if site=='Jingdong':
            break
else:
    print('遍历完毕')
```

终端中输出如下信息：

```
Baidu
Google
Jingdong
```

由于在遍历到"Jingdong"时，产生了break，因此遍历终端，没有输出"Taobao"和else子句。

3. while语句

while语句是同时包含条件和循环语句，用于根据条件的结果判断是否执行代码块，用法：

```
while condition:
    statements
else:
    statements
```

while语句可以单独使用，也可以结合else子句，与if语句相同，只有当condition为False时，才会执行else中的语句。

```
a=3
while a:
    print(a)
    a-=1
else:
    print('遍历完毕')
```

终端中输出如下信息：

```
3
2
1
遍历完毕
```

当a减小为0时，while的condition为False，所以执行了else子句。

4. break和continue

break语句和C语言中的类似，用于跳出最近的for或while循环。

continue语句也借鉴自C语言，表示继续执行循环的下一次迭代。

比较表1-1-2所示的执行结果就可以发现以下结果：

break将整个循环中止了，剩余的循环不再执行；

continue提前结束本次循环,后面的代码不再执行,直接进入下一次循环。

表1-1-2 执行结果对比表

break	continue

n = 5
while n:
    n -= 1
    if n == 2:
        break
    print(n)
print('循环结束。')
	n = 5
	while n:
	    n -= 1
	    if n == 2:
	        continue
	    print(n)
	print('循环结束。')

4 3 循环结束。	4 3 1 0 循环结束。

任务二 掌握Python数据类型

本任务是学习基础的Python数据类型,并掌握常用的方法。

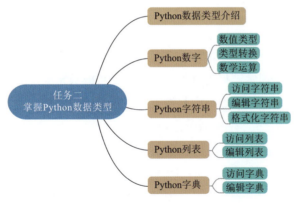

一、Python数据类型介绍

与其他语言不同,Python中数据类型指的不是变量。

Python中的变量不需要提前声明,在使用前必须赋值,赋值以后该变量才会被真正创建。

在Python中,变量就是变量,它没有类型,所说的"类型"是变量所指向的内存地址中对象的类型。

等号(=)用来给变量赋值;

等号(=)运算符左边是一个变量名,可以是符合Python语法的任意名字，甚至是中文；

等号（=）运算符右边是存储在变量中的值。

Python中的数据类型有多种，常用的基础类型有number（数字）、string（字符串）、bool（布尔类型）、list（列表）、tuple（元组）和dictionary（字典）。

bool比较简单，仅包含True和False；tuple可以认为是list的不可变形式，基本用法与list相似，所以不再过多解释。下面主要对另外四种基础类型进行讲解。

二、Python数字

Python数字用于存储数值。

Python数字是不可变的数据类型，这就意味着如果改变数字的值，将重新分配内存空间。

```
>>> var1=66
>>> var2=88
>>> var1, var2
    (66, 88)
```

1. 数值类型

Python支持三种不同的数值类型：

整型（int）：通常被称为整型或整数，是正或负整数，不带小数点。

浮点型（float）：浮点型由整数部分与小数部分组成，浮点型也可以使用科学计数法表示（2.5e2=2.5×10^2=250）。

复数（complex）：复数由实数部分和虚数部分构成，可以用a+bj，或者complex(a,b)表示，复数的实部a和虚部b都是浮点型。

2. 类型转换

不同的数值类型可以使用相应函数进行转换：

int(x)将x转换为一个整数。

float(x)将x转换为一个浮点数。

complex(x)将x转换为一个复数，实数部分为x，虚数部分为0。

complex(x, y)将x和y转换为一个复数，实数部分为x，虚数部分为y。x和y是数字表达式。

3. 数学运算

Python数字同样可以直接进行数学运算。

假设变量a=5，变量 b=2，运算符对照表见表1-1-3。

表1-1-3　运算符对照表

运算符	描述	实例
+	加，两个对象相加	a+b输出结果7
-	减，得到负数或是一个数减去另一个数	a-b输出结果3
*	乘，两个数相乘或是返回一个被重复若干次的字符串	a*b输出结果10
/	除，x除以y	b/a输出结果0.4
%	取余，返回除法的余数	b%a输出结果2
**	幂，返回x的y次幂	a**b为5的2次方，25

//	取整除，往小的方向取整数	>>> 9//2
		4
		>>> -9//2
		-5

Python还提供大量的数学函数和三角函数等，具体说明请查看官网文档。

三、Python字符串

Python字符串（string，简写为str），用于存储字符串，是Python中最常用的数据类型。可以使用''或""创建字符串。

Python字符串是不可变的数据类型，这就意味着如果改变字符串的值，将重新分配内存空间。

```
>>> var1 = 'Hello TQD!'
>>> var2 = 'Hello Python!'
>>> var1, var2
('Hello TQD!', 'Hello Python!')
```

1. 访问字符串

Python字符串支持的操作见表1-1-4。

表1-1-4　访问字符串

字符串操作	描述	实例	
+	连接符（同类型）	>>> 'he'+'llo' 'hello'	
*	重复操作	>>> 'hello'*3 'hellohellohello'	
[i]	索引。下标从0开始，负数表示从右向左	>>> 'hello'[1] 'e'	>>> 'hello'[-1] 'o'

[start:end:step]	切片，第一个':'不可省，其他均可省。三个参数均可是负数，表示从右向左	>>> 'hello'[:-1]
		'hell'
		>>> 'hello'[::-1]
		'olleh'	

续表

字符串操作	描述	实例
in	成员运算符True/False	>>> 'h' in 'hello' True
not in	非成员运算符True/False	>>> 'a' not in 'hello' True
del	删除字符串。str不可变，所以无法删除索引或切片，只能删除整个str	>>> s='hello' >>> del s

for...in...	遍历	>>> s = 'word'
		>>> for i in s:
		    print(i)
		w
		o
		r
		d

2. 编辑字符串

由于Python字符串是不可变的，所以编辑字符串指的是对原字符串进行索引、切片等，然后连接其他字符串，从而生成一个新的字符串。常用的编辑方法如下：

（1）大小写转换

```
str.capitalize() -> str
```

首字母大写。

```
>>> 'hello python'.capitalize()
'Hello python'
str.swapcase() -> str
```

大小写互换。

```
>>> 'Hello Python'.swapcase()
'hELLO pYTHON'
str.lower() -> str
```

全部小写。

```
>>> 'HELLO PYTHON'.lower()
'hello python'
str.upper() -> str
```

全部大写。

```
>>> 'hello python'.upper()
'HELLO PYTHON'
```

```
str.title() -> str
```

标题化,每个单词首字母大写。

```
>>> 'hello python'.title()
'Hello Python'
```

(2)查找与替换

```
str.find(sub[, start[, end])) ->int
```

查找sub,返回索引;若sub不存在,返回-1。

```
>>> 'hello python'.find('o')
4
>>> 'hello python'.find('a')
-1
str.rfind(sub[, start[, end])) -> int
```

从右侧查找sub,返回索引。

```
>>> 'hello python'.rfind('o')
10
str.index(sub[, start[, end]) -> int
```

查找sub,返回索引。与find的区别是,若sub不存在,index将报错。

```
>>> 'hello python'.index('o')
4
>>> 'hello python'.index('a')
Traceback (most recent call last):
  File "<pyshell#10>", line 1, in <module>
    'hello python'.index('a')
ValueError: substring not found
str.rindex(sub[, start[, end])) -> int
```

查找sub,返回索引。

```
>>> 'hello python'.rindex('o')
10
str.count(sub, start=None, end=None) -> int
```

统计sub出现的次数。

```
>>> 'hello python'.count('l')
2
str.replace(old, new [, max]) ->str
```

替换字符。

```
>>> 'hello python'.replace('p', 'P')
'hello Python'
```

(3) 填充与对齐

```
str.center(width, fillchar=' ') ->str
```

居中。

```
>>> 'TQD'.center(11)
'    TQD    '
>>> 'TQD'.center(11, '~')
'~~~~TQD~~~~'
str.ljust(width, fillchar=' ') ->str
```

左对齐。

```
>>> 'TQD'.ljust(11)
'TQD        '
str.rjust(width, fillchar=' ') ->str
```

右对齐。

```
>>> 'TQD'.rjust(11)
'        TQD'
```

(4) 移除与匹配

```
str.strip([chars]) ->str
```

移除开头和结尾空格或属于chars的字符。

```
>>> '  Hello python  '.strip()
'Hello python'
str.lstrip([chars]) ->str
```

移除开头空格或属于chars的字符。

```
>>> '  Hello python  '.lstrip()
'Hello python  '
str.rstrip([chars]) ->str
```

移除结尾空格或属于chars的字符。

```
>>> '  Hello python  '.rstrip()
'  Hello python'
```

(5) 连接与分割

```
str.join(iterable) ->str
```

连接iterable的所有元素,以str分割。

```
>>> ''.join(('Hello', 'TQD'))
'HelloTQD'
```

```
>>> ', '.join(('Hello', 'TQD'))
'Hello, TQD'
str.split(sep=None, maxsplit=-1) ->list
```

以sep分割字符串。

```
>>> 'Hello Python! Hello World!'.split(' ')
['Hello', 'Python!', 'Hello', 'World!']
str.rsplit(sep=None, maxsplit=-1) ->list
```

以sep分割字符串，从右向左。

```
>>> 'Hello Python! Hello World!'.rsplit(' ', 1)
['Hello Python! Hello', 'World!']
```

3. 格式化字符串

在程序中输出字符串的时候，经常需要随着实际情况进行调整，例如字符串中的种类、时间、数量、姓名、年龄等。

对每个需要输出的字符串，创建一一对应的变量，是一个非常烦琐的工作，这时就可以考虑使用格式化字符串。

```
'Python课程将于10:00开始，参与的学生总人数是22人。'
'数学课程将于11:00开始，参与的学生总人数是30人。'
'英语课程将于14:00开始，参与的学生总人数是28人。'
```

首先创建一个模板字符串，在使用时，根据实际情况进行格式化并输出。

```
'{}课程将于{}开始，参与的学生总人数是{}人。'
```

在Python中，字符串中的花括号"{}"可用于占位符，表示这是一个变量，可以进行格式化。格式化的方式有三种：

%格式化：这是在Python 2时代使用的格式化方式；

format格式化：字符串的format()方法，自由度非常高；

f-string格式化：f-string是标注了'f'或'F'前缀的字符串，可以直接在"{}"中使用上文中出现过的变量。

由于%格式化已经不推荐使用，下面主要介绍format格式化和f-string格式化。

1）format格式化

```
str.format(*args, **kwargs)
```

（1）序号格式化

按数字序号来替换str中的'{}'占位字段。

```
>>> '{0}{2}{1}'.format('hello', 'ld', 'wor')
'hello world'
```

如果序号是0,1,2,…的顺序,则可以省略,但此时'{}'与format参数个数必须相同。

```
>>> '{} {}'.format('hello', 'world')
hello world
>>> template = '{}课程将于{}开始,参与的学生总人数是{}人.'
>>> template.format('Python', '10: 00', 22)
'Python课程将于10:00开始,参与的学生总人数是22人.'
>>> template.format('数学', '11: 00', 30)
'数学课程将于11:00开始,参与的学生总人数是30人.'
>>> template.format('英语', '14: 00', 28)
'英语课程将于14:00开始,参与的学生总人数是28人.'
```

（2）关键字格式化

按照关键字替换'{}'占位字段。

```
>>> template = '{subject}课程将于{time}开始,参与的学生总人数是{students}人.'
>>> template.format(subject='Python', time='10: 00', students=22)
'Python课程将于10:00开始,参与的学生总人数是22人.'
>>> template.format(subject='数学', time='11: 00', students=30)
'数学课程将于11:00开始,参与的学生总人数是30人.'
>>> template.format(subject='英语', time='14: 00', students=28)
'英语课程将于14:00开始,参与的学生总人数是28人.'
```

（3）数字格式化

格式化显示字符串中的数字。

```
>>> '{:.2f}'.format(3.1415926)     # 保留两位小数
'3.14'
>>> '{:>5}'.format(3.14)           # 右对齐,总宽度5个字符
' 3.14'
>>> '{:~>5}'.format(3.14)          # 右对齐,总宽度5个字符,使用~填充
'~3.14'
>>> '{:>5.2f}'.format(3.1415926)   # 右对齐,总宽度5个字符;保留两位小数
' 3.14'
```

2）f-string格式化

相比较format，f-string格式化可以直接在'{}'中执行表达式。

```
>>> f'{3>2}'
'True'
>>> food = ['西瓜', '葡萄', '香蕉']
>>> f'{food[0]}'
'西瓜'
>>> class Car:
        color = 'red'
>>> car = Car()
>>> f'{car.color}'
'red'
```

f-string同样支持对字符串中的数字进行格式化。

```
>>> value = 3.1415926
>>> f'{value:.2f}'
'3.14'
>>> f'{value:>5.2f}'
' 3.14'
>>> f'{value:~>5.2f}'
'~3.14'
```

四、Python列表

Python列表（list），是一个序列，用于存储多个数据，种类没有限制，可以相同，也可以不同。可以使用中括号"[]"创建列表。

Python列表是可变的数据类型，可以对其中的元素进行增加、删除或修改。

```
>>> list1 = ['C/C++', 'Java', 'Python', 'Go']
>>> list2 = ['小明', 20, 88]
```

1. 访问列表

Python列表支持的操作见表1-1-5。创建一个变量，>>> l = [22,33,44,55]。

表1-1-5　访问列表

列表操作	描 述	实 例	
+	连接符（同类型）	>>> [1,2,3]+[7,8] [1, 2, 3, 7, 8]	
*	重复操作	>>> [1,2]*3 [1, 2, 1, 2, 1, 2]	
[i]	索引。下标从0开始，负数表示从右向左	>>> l[0] 22	>>> l[-1] 44

[start:end:step]	切片，第一个':'不可省，其他均可省。三个参数均可是负数，表示从右向左	>>> l[:2]
		[22, 33]
		>>> l[::2]
		[22, 44]	

in	成员运算符True/False	>>> 66 in l False	
not in	非成员运算符True/False	>>> 44 not in l True	

del	删除元素或切片 del list[i]
	删除整个列表 del list
		>>> del l[2]
		>>> l
		[22, 33, 55]	

续表

列表操作	描　述	实　例
for...in...	遍历	>>> for i in l: 　　print(i) 22 33 44 55

2. 编辑列表

Python列表是可变类型数据，增加、删除或修改是对列表本身进行修改的。

（1）查找与统计

```
list.index(x[, start[, end]]) ->int
```

查找x的索引。

```
>>> list1 = ['C/C++', 'Java', 'Python', 'Go']
>>> list1.index('Python')
2
list.count(x) ->int
```

统计某个元素在列表中出现的次数。

```
>>> list1.count('Python')
1
```

（2）插入与扩展

```
list.append(x) ->None
```

追加，在列表末尾添加元素x。

```
>>> list1.append('C#')
>>> list1
['C/C++', 'Java', 'Python', 'Go', 'C#']
list.insert(i, x) ->None
```

插入，在索引i位置插入元素x。

```
>>> list1.insert(3, 'JavaScript')
>>> list1
['C/C++', 'Java', 'Python', 'JavaScript', 'Go', 'C#']
list.extend(iterable) ->None
```

扩展，扩展iterable到列表尾部。

```
>>> list1.extend(('VB', 'SQL'))
>>> list1
['C/C++', 'Java', 'Python', 'JavaScript', 'Go', 'C#', 'VB', 'SQL']
```

(3) 删除与清空

```
list.pop([i]) ->object
```

基于索引删除元素。

```
>>> list1.pop()
'SQL'
>>> list1.pop(-2)
'C#'
list.remove(x) ->None
```

基于值删除元素，只删除第一个。

```
>>> list1.remove('Go')
>>> list1
['C/C++', 'Java', 'Python', 'JavaScript', 'VB']
list.clear() ->None
```

清空列表。

```
>>> list1.clear()
>>> list1
[]
```

(4) 排序

```
list.sort(*, key=None, reverse=False) ->None
```

对列表排序。

```
>>> list1 = ['C/C++', 'Java', 'Python', 'Go']
>>> list1.sort()
>>> list1
['C/C++', 'Go', 'Java', 'Python']
list.reverse() ->None
```

翻转列表中的元素，不考虑大小。

```
>>> list1.reverse()
>>> list1
['Python', 'Java', 'Go', 'C/C++']
```

五、Python字典

Python字典（dict），同样属于序列，是一个键值对的容器。可以使用花括号"{}"创建字典。

Python字典是可变的数据类型，可以对其中的键值对进行增加、删除或修改。

```
>>> dict1 = {'name':'li', 'age':18, 'score':90}
```

1. 访问字典

Python字典中每个元素是一个键值对,支持的操作见表1-1-6。创建一个变量,>>> d = {'a':1, 'b':2}。

表1-1-6　访问字典

字典操作	描　　述	实　　例

[key]	按key访问或赋值,dict不支持索引和切片	>>> d['a'] = 5
		>>> d['a']
		5

in	成员(key)运算符True/False	>>> 'a' in d True
not in	非成员(key)运算符True/False	>>> 'd' not in d True

del	删除某个元素 del dict[key]
	删除整个字典 del dict
		>>> del d['a']
		>>> d
		{'b': 2}

for...in...	迭代	>>> for i in d: 　　　print(i) a b

与字符串和列表不同的是,字典不支持索引和切片。

2. 编辑字典

Python字典是可变类型数据,增加、删除或修改是对字典本身进行修改。

(1) 查找与统计

```
dict.get(key[, default]) ->value
```

返回key的value,若不存在则返回default。

```
>>> dict1 = {'name':'li', 'age':18, 'score':90}
>>> dict1.get('name')
'li'
dict.setdefault(key[, default]) ->value
```

返回key的value,若不存在则加入key:default键值对,并返回default。

```
>>> dict1.setdefault('subject', 'Python')
'Python'
>>> dict1
{'name': 'li', 'age': 18, 'score': 90, 'subject': 'Python'}
dict.items() -> dict_items[_KT, _VT]
```

返回键值对(元组)构成的视图,可以迭代,但不可以索引切片。如果有需要,可用list展开。

```
>>> dict1.items()
dict_items([('name', 'li'), ('age', 18), ('score', 90), ('subject',
'Python')])
    dict.keys() -> dict_keys[_KT, _VT]
```

返回keys构成的视图，可以迭代，但不可以索引切片。如果有需要，可用list展开。

```
>>> dict1.keys()
dict_keys(['name', 'age', 'score', 'subject'])
    dict.values() -> dict_values[_KT, _VT]
```

返回values构成的视图，可以迭代，但不可以索引切片。如果有需要，可用list展开。

```
>>> dict1.values()
dict_values(['li', 18, 90, 'Python'])
```

（2）删除与清空

```
dict.pop(key[,default]) ->value
```

删除键值对，返回value。若key不存在也没给定default，将触发KeyError。

```
>>> dict1.pop('age')
18
>>> dict1
{'name': 'li', 'score': 90, 'subject': 'Python'}
    dict.popitem() ->tuple
```

删除最后一个键值对，并返回其元组。

```
>>> dict1.popitem()
('subject', 'Python')
>>> dict1
{'name': 'li', 'score': 90}
    dict.clear() ->None
```

清空字典内所有元素。

```
>>> dict1.clear()
>>> dict1
{}
```

任务三　学习Python函数和类

本任务是学习Python函数和类，并掌握常用的方法。

一、Python函数

函数是封装好的，可重复使用的，用来实现单一或相关联功能的代码段。函数能提高程序的模块性和代码的重复利用率。

1. 内置函数

Python解释器内置了很多函数，无须导入就可以直接使用，见表1-1-7。对于每个内置函数的用法详见官网。

表1-1-7 内置函数对照表

A	E	L	R
abs()	enumerate()	len()	range()
aiter()	eval()	list()	repr()
all()	exec()	locals()	reversed()
any()			round()
anext()	F	M	
ascii()	filter()	map()	S
	float()	max()	set()
B	format()	memoryview()	setattr()
bin()	frozenset()	min()	slice()
bool()			sorted()
breakpoint()	G	N	staticmethod()
bytearray()	getattr()	next()	str()
bytes()	globals()		sum()
		O	super()
C	H	object()	
callable()	hasattr()	oct()	T
chr()	hash()	open()	tuple()
classmethod()	help()	ord()	type()
compile()	hex()		
complex()		P	V
	I	pow()	vars()
D	id()	print()	
delattr()	input()	property()	Z
dict()	int()		zip()
dir()	isinstance()		
divmod()	issubclass()		__import__()
	iter()		

2. 自定义函数

除了使用内置函数外，还可以自定义函数，实现自己需要的功能。

（1）函数定义

```
def 函数名([参数]):
    # 内部代码
    return 表达式
```

Python使用关键字def定义函数；

函数名是必须提供的，可以是符合Python语法的任意名称，甚至是中文；

函数名后面必须有括号，用于定义参数；若无须使用参数，可以省去；

括号后面必须紧跟冒号"："，表示代码块的开始；

函数内部的第一级缩进必须相同，表示这个代码块是一个整体；

函数结尾应当使用return返回返回值，若没有return语句，Python将默认为return None。

```
def 功能():
    return 1

print(功能())              # 输出返回值，即1
```

若函数需要使用参数，可以在函数名后面的括号中定义，这里的参数名就是变量名。参数主要有两种，分别是位置参数和关键字参数。

（2）位置参数

位置参数又称必传参数、顺序参数，是最重要的，也是必须在调用函数时明确提供的参数。位置参数必须按先后顺序，一一对应，个数不多不少地传递。

注意：Python在做函数参数传递的时候不会对数据类型进行检查，理论上传什么类型都可以，但是在实际运算的时候如果发现数据类型错误，将会弹出异常。

```
def fun(a, b, c):
    return a + b + c

print(fun(5, 2, 1))      # 8
```

（3）关键字参数

调用函数时，所有的参数都是按位置（顺序）传入函数内部的，如果传递顺序与使用顺序不匹配，将出现各种各样的问题，甚至是崩溃。

使用关键字传参可以避免这种情况，不需要按照顺序。关键字传参就是使用"参数名=value"的形式，传入参数。

```
def fun(a, b):
    return a ** b


print(fun(2, 3))           # 8
print(fun(3, 2))           # 9
print(fun(b=3, a=2))       # 关键字传参,与顺序无关,8
```

（4）默认值参数

在函数定义时，还可以给某个参数提供一个默认值。默认值参数也属于关键字参数。在调用时，可以给关键字参数传递一个自定义的值，也可以不传参，即使用默认值。

```
def fun(a=2, b=3):
    return a ** b

print(fun())               # 8
print(fun(5, 2))           # 25
```

当位置参数和关键字参数组合使用时，Python语法要求关键字参数必须在位置参数后面，否则将触发异常。

```
def fun(a, b=3):
    return a ** b

print(fun(5))              # 125
```

关键字参数放在位置参数前面，触发异常。

```
def fun(a=3, b):
    return a ** b
SyntaxError: non-default argument follows default argument
```

二、Python类

Python是一种面向对象的语言，Python中的类提供了面向对象编程的所有基本功能。类的继承机制，允许派生类可以覆盖基类中的任何方法，方法中可以调用基类中的同名方法。

1. 类定义

```
class 类名:
    def 函数名(self, [参数]):
        # 内部代码
        return 表达式
    ...
```

Python使用关键字class定义类；

类名必须提供，且推荐首字母大写；

如果继承了其他类，类名后面可以添加括号，将基类放在括号中；

类中可以定义函数，称为实例方法，必须接受一个参数self，表示实例对象自身；

定义在函数之外的变量是类变量，与普通变量相同；

定义在函数之内的变量是实例属性，必须以self.变量表示。

```
class Car:
    def __init__(self, person, color, seats):
        self.person = person
        self.color = color
        self.seats = seats

    def say(self):
        a = 2
        print(f'{self.person}的汽车是{self.color}，共有{self.seats}个座位')

    def get_color(self):
        return self.color

    def set_color(self, color):
        self.color = color
```

2. 实例化

如上例所示，Car是一个类，指向内存地址；而Car()是执行，是对类的实例化。若使用参数，实例化时需要将参数传入。

```
mycar = Car('小明', '红色', '4')
```

3. 构造方法

构造方法是类中的一个特殊的实例方法，名字固定为"__init__"会在类实例化时首先完成，可以理解为初始化。实例化时传入的参数，都会被送入构造方法。

```
def __init__(self, …):
    代码块
```

所有的实例方法，包括构造方法，都必须接收一个self参数，表示实例对象本身；

实例属性必须以self.变量的形式表示。

4. 实例方法

实例方法是类中的功能函数，可以是内部调用的，也可以对外开放。若实例方法无法满足需求，还是继承重写。

```
def name(self, …):
    代码块
```

除了self参数外，实例方法还可以添加其他额外参数。但要注意，这些额外参数并没有在实例化时传入，所以在调用实例方法时，还需将额外的参数传入进去。

5. 类的继承

继承机制经常用于创建和现有类功能类似的新类，又或是新类只需要在现有类基础上添加或修改一些属性或方法，但又不想直接将现有类代码复制给新类。也就是说，通过使用继承这种机制，可以轻松实现类的重复使用。

```
class Parent :
    pass


class Child(Parent):
    pass
```

① 子类可以直接调用父类属性和方法。

② 子类可以定义新的属性和方法。

③ 子类可以重写父类原有的属性和方法。

任务四　掌握Python扩展库

本任务是了解Python扩展库，并掌握常用的标准库和第三方库。

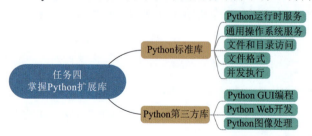

一、Python标准库

标准库是Python集成的扩展库，随着Python客户端的安装而安装到本地。由于Python标准库数量超过240个，所以本书仅介绍一些常用的标准库。

1. Python运行时服务

sys　提供系统相关的参数和函数。这些变量可能被解释器使用，也可能由解释器提供。这些函数会影响解释器。

2. 通用操作系统服务

（1）os

提供多种操作系统接口功能。如果只是读写一个文件，请使用open()函数；如果想操作文件路径，请参阅os.path模块；如果想读取通过命令行给出的所有文件中的所有行，请参阅 fileinput模块；为了创建临时文件和目录，请参阅tempfile模块，对于高级文件和目录处理，请参阅shutil模块。

（2）time

提供时间的访问和转换，相关功能还可以参阅datetime和calendar模块。

3. 文件和目录访问

（1）os.path

提供常用路径操作功能。不同的操作系统具有不同的路径名称约定，因此os.path标准库集成了转换功能，输出的路径始终都是适合本机操作系统的路径。

（2）shutil

提供高阶文件操作功能，特别是复制和删除。

4. 文件格式

（1）configparser

INI文件是initialization file的缩写，即初始化文件，是Windows的系统配置文件所采用的存储格式，统管Windows的各项配置，configparser模块提供了INI文件解析的功能。

（2）csv

csv（comma separated values）格式是电子表格和数据库中最常见的输入、输出文件格式。csv模块提供对CSV文件的读写功能。

5. 并发执行

（1）threading

提供基于线程的并行。由于存在全局解释器锁，同一时刻只有一个线程可以执行（虽然某些性能导向的库可能会去除此限制）。因此，如果想充分利用多核计算机的性能,推荐使用 multiprocessing。如果同时运行的任务是I/O密集型的,则多线程仍然是一个较合适的选择。

(2) multiprocessing

提供基于进程的并行。通过使用子进程而非线程有效地绕过了全局解释器锁。因此,multiprocessing模块允许程序员充分利用计算机的多核性能。

(3) subprocess

提供子进程管理的功能,可以连接它们的输入、输出和错误管道,并获取它们的返回值。

二、Python第三方库

第三方库是Python客户端没有集成的扩展库,可以通过pip命令安装使用。

1. Python GUI编程

GUI是graphical user interface(图形用户接口)的缩写。Python自带的tkinter模块可以快速开发简单桌面应用。第三方库如PyQt、PySide、PySimpleGUI、Kivy、wxPython等等,都可以用来进行GUI编程。这里推荐使用PySide。

PySide是跨平台应用程序框架Qt的Python绑定,Qt是跨平台C++图形可视化界面应用开发框架,自推出以来深受业界盛赞。PySide由Qt公司自己维护,最新版本是PySide6,允许用户在Python环境下利用Qt开发大型复杂GUI。PySide6支持LGPL协议,可以使用动态链接的形式开发闭源的程序,可以以任何形式(商业的、非商业的、开源的、非开源的等)发布应用程序。

使用如下命令安装PySide6:

```
pip install PySide6
```

如下所示,是一个非常简单的按钮程序。

```python
from PySide6.QtWidgets import QWidget, QApplication, QPushButton
import sys

class Mydemo(QWidget):
    def __init__(self):
        super().__init__()
        btn = QPushButton('按钮', self)
        btn.move(50, 50)
        btn.setStyleSheet("QPushButton:pressed {color:blue}")

if __name__ == '__main__':
```

```
app = QApplication(sys.argv)
win = Mydemo()
win.show()
sys.exit(app.exec())
```

运行后，就可以看到GUI界面了，单击按钮时，文本将变为蓝色，如图1-1-10所示。

2. Python Web开发

典型的Web应用开发框架有Django、Flask、Pyramid，可以选择自己感兴趣的学习。推荐使用Flask进行简单的Web开发。

图1-1-10　简单GUI界面

Flask是一个使用Python编写的轻量级Web应用框架。它被称为微框架（microframework），"微"并不是意味着把整个Web应用放入一个Python文件，微框架中的"微"是指Flask旨在保持代码简洁且易于扩展，Flask框架的主要特征是核心构成比较简单，但具有很强的扩展性和兼容性，程序员可以使用Python语言快速实现一个网站或Web服务，用户可以根据需要进行扩展。

使用如下命令安装Flask：

```
pip install flask
```

如下所示，是一个非常简单的Web程序。

```
from flask import Flask

app = Flask(__name__)


@app.route('/')
def hello_world():
    return 'Hello World!'


if __name__ == '__main__':
    app.run()
```

运行后，在浏览器中访问http://127.0.0.1:5000就可以看到显示的内容了，如图1-1-11所示。

3. Python图像处理

在Python中不可避免地要进行图像的处理，如读、写、识别等，最常用的有两个库OpenCV

图1-1-11　简单Web服务

和Pillow。这里推荐使用OpenCV。

OpenCV是由英特尔公司资助的开源计算机视觉库，可以运行在Linux、Windows、Android和MacOS操作系统上，提供Python接口，可以非常方便地实现图像处理和计算机视觉方面的很多通用算法。

使用如下命令安装OpenCV：

```
pip install opencv-python
```

如下所示，是一个非常简单的取色器程序。

```
import cv2

def mouseColor(event, x, y, flags, param):
    if event == cv2.EVENT_LBUTTONDOWN:
        print('HSV:', hsv[y, x])

img = cv2.imread('2.jpg')
hsv = cv2.cvtColor(img, cv2.COLOR_BGR2HSV)
cv2.imshow("Color Picker", img)
cv2.setMouseCallback("Color Picker", mouseColor)
if cv2.waitKey(0) == ord('q'):
    cv2.destroyAllWindows()
```

运行程序后，单击图像上的任意位置，将输出该位置的HSV分量，如图1-1-12所示。

图1-1-12 取色器

思考与总结

1. 不同的项目需要的扩展库是不同的，如何维护多个项目？

2. pip还有哪些常用方法？

3. 为了提高项目代码的效率，建议将特定的代码写成函数或者类，通过调用实现复用。

项目二

学习环境搭建

 学习目标

1. 掌握Ubuntu系统的安装与基本使用方式。
2. 掌握ROS2的安装，了解ROS2的发展历程。
3. 掌握Gazebo的安装，并与ROS2集成。

任务一　了解Ubuntu系统

本任务目的是对Ubuntu系统有一定的了解，学习如何安装Ubuntu系统，并熟悉常用系统命令。

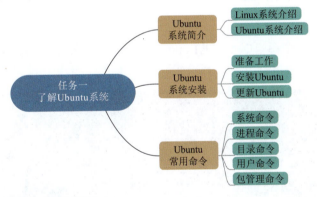

一、Ubuntu系统简介

1. Linux系统介绍

Linux系统由Linus Torvalds（莱纳斯·托瓦尔兹）于1991年发布在新闻组的内核发展而来。由于它在发布之初就免费和自由传播，支持多用户、多任务及多线程，且兼容POSIX标准，它支持运行当时主流系统Unix的一些工具软件，吸引了众多的使用者和开发者，逐渐发展壮大至今。

由于Linux内核本身是开源的，基于Linux内核搭配各种各样系统管理软件或应用工具软件，从而组成一套完整可用的操作系统，如图1-2-1所示。完整的

Linux系统就如同汽车，Linux内核构成了最为关键的引擎，不同的发行版就类似使用相同引擎的不同车型。

人们制作发行版通常用于特定的用途，侧重点有所不同，因而Linux发行版可谓百花齐放。其中，以Debian、Suse及Fedora的发行版最为常见，如图1-2-2所示。

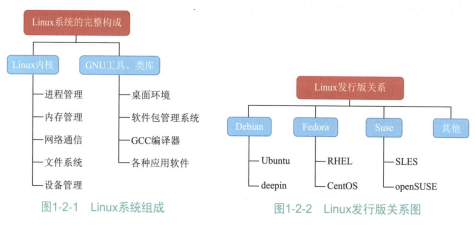

图1-2-1　Linux系统组成

图1-2-2　Linux发行版关系图

对于初次接触Linux的用户，建议只要了解到这些都是Linux系统，而且目前开发主机要用的是Ubuntu即可。

2. Ubuntu系统介绍

Ubuntu是一个源自Debian的Linux发行版。Ubuntu 22.04桌面如图1-2-3所示，请自行在搜索引擎中搜索下载网址。

图1-2-3　Ubuntu 22.04桌面

Ubuntu专注于用户和可用性，有以下特点：

（1）完整的桌面系统

Ubuntu为运营组织、学校、家庭或企业提供了所需的一切。预装了所有必要的应用程序，例如办公套件、浏览器、电子邮件和多媒体应用等。Ubuntu软件中心提供了成千上万的游戏和应用程序。

（2）开源

Ubuntu一直是免费下载、使用和分享。

（3）安全

Ubuntu是最为安全的操作系统之一，其内建了防火墙和病毒保护软件。

（4）可访问

计算用于所有人，不论国籍、性别等。Ubuntu被完整地翻译成50多种语言，且包含了必要的辅助技术。

（5）系统界面效果

Ubuntu已获得高清触摸屏，桌面UI缩放，以及触摸板手势的支持。22.04 LTS更新了标志性的Yaru主题，具有系统范围的深色风格偏好支持、强调色和迄今为止最大的社区壁纸集。

（6）广泛的硬件支持

Canonical与戴尔、联想和惠普紧密合作，以证明Ubuntu可在最广泛的笔记本计算机和工作站上使用。可以提供前所未有的无缝Ubuntu体验，以及比以往更多的硬件选择。

Ubuntu不仅只适用于桌面计算机，也被广泛地使用在世界各地的数据中心中，为各种各样的服务器提供支持，并且还是云计算中最流行的操作系统。

Ubuntu文件目录如图1-2-4所示。

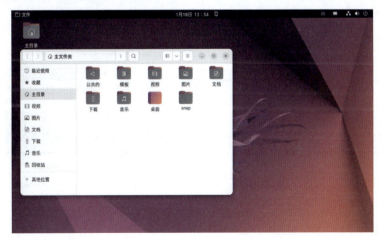

图1-2-4　文件目录

Ubuntu系统设置如图1-2-5所示。

图1-2-5　Ubuntu系统设置

二、Ubuntu系统安装

本书基于最新的长期支持的Ubuntu版本，即Ubuntu 22.04 LTS。推荐将Ubuntu系统直接安装到物理真机上，若安装到虚拟机中，可能会影响仿真的性能。

1. 准备工作

首先下载Ubuntu镜像文件，下载地址请自行在搜索引擎中搜索获得。

要安装Ubuntu Desktop，还需要将下载的ISO写入U盘以创建安装介质（启动盘）。这与复制ISO不同，需要一些定制软件。本书基于Windows系统，使用官方推荐的Rufus软件，制作U盘启动盘。

Rufus下载地址请自行在搜索引擎中搜索获得。

Rufus软件下载后无须安装，Windows 7或更高版本（32/64位均可）即开即用。

然后制作U盘启动盘，Rufus设置如图1-2-6所示，单击"开始"按钮，等待制作完成。

接着压缩硬盘，如果直接覆盖Windows系统，无须此步骤。

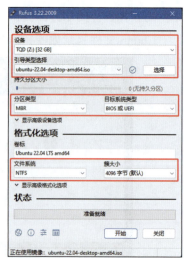

图1-2-6　Rufus设置

压缩硬盘，分隔出用于安装Ubuntu的硬盘，大小根据自己的需要设置。Ubuntu本身要占用10 GB左右，所以建议压缩30 GB以上，如图1-2-7所示。

注意：压缩出来的硬盘空间不要分配，Ubuntu会自动安装到未分配的硬盘区域。如果分配了，就需要在安装时重新设置。

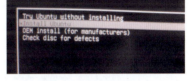

图1-2-7　压缩硬盘

最后还需要设置BIOS，将第一个启动方式设置为USB启动。

不同品牌的计算机进入BIOS的方式不同，请自行搜索。设置完毕后，保存并重启计算机。

2. 安装Ubuntu

重启计算机后将进入U盘启动模式，选择第二项Install Ubuntu，如图1-2-8所示。

图1-2-8　安装Ubuntu

语言选择"中文"，键盘选择"英语（美国）"，更新和其他软件中选择"正常安装"。

在图1-2-9所示安装类型中，如果安装双系统，就选择第一项"安装Ubuntu，与Windows Boot Manager共存"，如果覆盖Windows，就选择第二项"清除整个磁盘并安装Ubuntu"。

图1-2-9　安装类型

单击"继续"按钮，在最后设置用户名和密码，即可完成安装。

3. 更新Ubuntu

首先进入Ubuntu系统。

Ubuntu系统安装完成后，重启计算机，在启动界面会显示Ubuntu引导，如图1-2-10所示。

图1-2-10　Ubuntu引导

第一项是启动Ubuntu系统，第三项是启动Windows系统，默认启动Ubuntu系统，倒计时10s，可以用键盘上的箭头【↑】或【↓】或鼠标选择启动Windows系统。

首先打开"软件和更新"，修改下载源为国内源，例如华为源、清华源等，如图1-2-11所示。

图1-2-11　修改下载源

打开软件更新器，更新系统。然后进行语言设置。

Ubuntu系统默认使用英文，如果有需要可以安装中文。

如图1-2-12所示，单击"1"，安装中文语言，然后在"2"中设置使用的语言为中文即可。

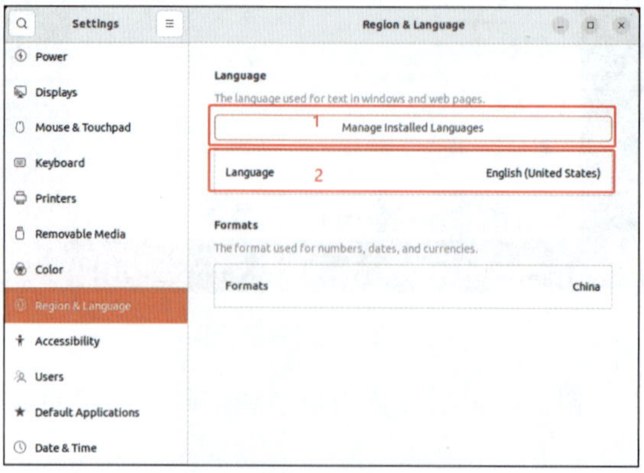

图1-2-12　安装中文语言

最后安装需要使用的软件。

Ubuntu内置了软件商店，可以根据自己的需要安装软件。这里推荐安装Visual Studio Code，简称VSCode，在Ubuntu系统软件商店中显示为code，用于以后的代码编辑，如图1-2-13所示。

图1-2-13　安装VSCode

三、Ubuntu常用命令

本书使用的Ubuntu系统包含图形化桌面，可以像Windows系统一样进行各项操作。但更常用的还是通过终端进行操作。在任意目录右击鼠标，选择"在终端中打开"命令，或使用快捷键【Ctrl+Alt+T】，即可启动终端，如图1-2-14所示。

Ubuntu的常用命令见表1-2-1～表1-2-5。

图1-2-14　右键菜单

表1-2-1　系统命令

命　　令	描　　述
shutdown	关机
poweroff	关机
reboot	重启

表1-2-2　进程命令

命　　令	描　　述
ps	显示当前进程
top	实时显示系统中各个进程的资源占用状况，类似于Windows的任务管理器
kill	终止指定的进程

表1-2-3　目录命令

命令	描述
pwd	显示工作路径
cd	切换shell工作目录
ls	显示当前工作目录中的内容
touch	修改文件的访问时间和修改时间，若文件不存在，会自动创建
mkdir	创建目录命令
rm	删除文件或目录

表1-2-4　用户命令

命令	描述
su	用户切换命令
sudo	普通用户临时具有root权限
passwd	修改密码，必须有root权限才可以修改

表1-2-5　包管理命令

命令	描述
pip	Python专用命令，用于安装第三方库
apt-get	用于从Ubuntu软件仓库中搜索、安装、升级、卸载软件

任务二　学习ROS2的安装

本任务目的是对ROS2有一定的了解，学习如何安装ROS2。

```
任务二           ROS2简介 ── ROS简介
学习ROS2的安装              └ ROS2与ROS的对比
                 ROS2安装 ── 安装ROS2
                           ├ 设置环境变量
                           └ 体验海龟模拟器
```

一、ROS2简介

1. ROS简介

ROS诞生于2007年，是一个适用于机器人编程的框架，如图1-2-15所示，这个框架把原本松散的零部件耦合在了一起，为它们提供了通信架构。ROS虽然称为操作系统，但并非Windows、Mac那样通常意义的操作系统，它只是连接了

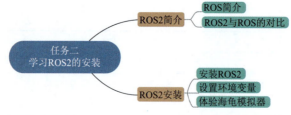

图1-2-15　ROS

操作系统和开发的ROS应用程序,所以它也算是一个中间件,基于ROS的应用程序之间建立起了沟通的桥梁,所以也是运行在Linux上的运行环境,在这个环境中,机器人的感知、决策、控制算法可以更好地组织和运行。

ROS主要具有以下特点:

(1) 分布式、点对点

ROS采用了分布式的网络框架,使用了基于TCP/IP的通信方式,实现了模块间点对点的松耦合连接,可以执行多种类型的通信。通过点对点的设计让机器人的进程可以分别运行,便于模块化修改和定制,提高了系统的容错能力。

(2) 支持多种语言

ROS支持多种编程语言。C++和Python是目前应用最广的ROS开发语言。此外,ROS还支持LISP、C#、Java、Octave等多种不同的语言。为了支持更多应用的移植和开发,ROS采用了一种语言中立的接口定义语言来实现各模块之间消息传送。通俗的理解就是,ROS的通信格式和用哪种编程语言无关,它使用的是自身定义的一套通信接口。

(3) 组件化工具包丰富

ROS采用组件化的方式将已有的工具和软件进行了集成,比如ROS中三维可视化平台Rviz。Rviz是ROS自带的一个图形化工具,可以方便地对ROS的程序进行图形化操作;ROS中常用的物理仿真平台Gazebo,在该仿真平台下可以创建一个虚拟的机器人仿真环境,还可以在仿真环境下添加一些需要的参数。

(4) 免费且开源

ROS具有一个庞大的开源社区ROS WIKI。ROS WIKI中的应用代码以维护者来分类,主要包含由Willow Garage公司和一些开发者设计、维护的核心库部分,以及不同国家的ROS社区组织开发和维护的全球范围的开源代码。当前使用ROS开发的软件包已经达到数千万个,相关的机器人已经多达上千款。此外,ROS遵从BSD协议,允许使用者修改和重新发布其中的应用代码,对个人和商业应用及修改完全免费。

2. ROS2与ROS的对比

ROS发布以来就被应用于各种各样的机器人,包括轮式机器人、腿式机器人、工业手臂、室外无人车辆、自动驾驶汽车、无人机、无人艇等。ROS已经从最初的学术研究项目,演变成一个工业界应用广泛的商业项目,已经成为机器人领域的事实标准。

随着应用场景的扩大,ROS最初设计时的局限性已经难以满足更多新需求，比如说实时性和网络延迟等，于是ROS2诞生了。

ROS2是在ROS的基础上设计开发的第二代机器人操作系统，可以帮助简化机器人开发任务，加速机器人落地的软件库和工具集。相比较ROS，ROS2有以下三个最显著的特点：

（1）架构的颠覆

在ROS的架构下，所有节点需要使用Master进行管理，也就是说要启动节点之前，必须向ROS_Master进行注册才可以使用。

ROS2使用基于DDS的Discovery机制，不需要使用roscore来启动Master，直接启动节点即可。

（2）API重新设计

ROS中的大部分代码都基于2009年2月设计的API。

ROS2重新设计了用户API，但使用方法类似，ROS2可以使用Python3。

（3）编译系统的升级

ROS使用rosbuild、catkin管理项目。

ROS2使用升级版的ament、colcon，见表1-2-6。

表1-2-6　ROS2更新计划

发 行 版	发 布 时 间	支 持 期 限
Humble Hawksbill	2022年5月23日	2027年5月
Galactic Geochelone	2021年5月23日	2022年12月9日
Foxy Fitzroy	2020年6月5日	2023年5月
Eloquent Elusor	2019年11月22日	2020年11月
Dashing Diademata	2019年5月31日	2021年5月
Crystal Clemmys	2018年12月14日	2019年12月
Bouncy Bolson	2018年7月2日	2019年7月
Ardent Apalone	2017年12月8日	2018年12月
beta3	2017年9月13日	2017年12月
beta2	2017年7月5日	2017年9月
beta1	2016年12月19日	2017年7月
alpha1～alpha8	2015年8月31日	2016年12月

二、ROS2安装

本书使用ROS2 Humble作为首选版本。安装过程需要连接网络，安装步骤如下。

1. 安装ROS2

首先更换国内源，比如华为源，安装过程会更便捷。

```
$ echo "deb [arch=$(dpkg --print-architecture)] https://repo.huaweicloud.com/ros2/ubuntu/(lsb_release -cs) main" | sudo tee /etc/apt/sources.list.d/ros2.list > /dev/null
```

添加密钥。

```
$ sudo apt install curl gnupg2 -y
$ curl -s https://gitee.com/clk_china/rosdistro/raw/master/ros.asc | sudo apt-key add -
```

更新系统。

```
$ sudo apt-get update
$ sudo apt-get upgrade
```

安装ROS2 Humble。

```
$ sudo apt install ros-humble-desktop
```

安装额外依赖。

```
$ sudo apt install python3-argcomplete -y
```

2. 设置环境变量

安装完毕后，将ros2命令加入环境变量，就可以直接使用ROS2相关命令了。

```
$ echo "source /opt/ros/humble/setup.bash" >> ~/.bashrc
```

编译配置文件，使其立即生效。

```
$ source ~/.bashrc
```

输入ros2。

```
$ ros2
```

终端中输出如下信息。

```
usage: ros2 [-h] [--use-python-default-buffering]
            Call 'ros2 <command> -h' for more detailed usage. …

ros2 is an extensible command-line tool for ROS 2.
…
```

说明ROS2 Humble已经安装完成了。

3. 体验海龟模拟器

海龟模拟器是ROS2内置的一个小游戏，如图1-2-16所示，下面使用命令来启动它：

```
$ ros2 run turtlesim turtlesim_node
```

图1-2-16　海龟模拟器

新开一个终端，启动海龟的控制节点，如图1-2-17所示。

```
$ ros2 run turtlesim turtle_teleop_key
```

图1-2-17　海龟控制节点

使用上下左右箭头移动海龟，使用【E】【R】【T】【D】【G】【C】【V】【B】旋转海龟，使用【F】停止旋转，使用【Q】退出。

任务三　学习Gazebo安装与集成

本任务目的是对Gazebo有一定的了解，学习如何安装Gazebo。

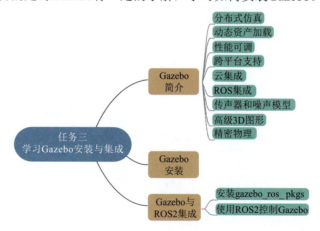

视　频

海龟模拟器体验

一、Gazebo简介

Gazebo是ROS和ROS2中最常用的机器人仿真平台，如图1-2-18所示。

其官方网址请自行在搜索引擎中搜索获得。

图1-2-18　Gazebo仿真图像

Gazebo是一个独立的应用程序，可以独立使用。主要有以下特点：

1. 分布式仿真

Gazebo支持使用多台服务器来提高性能。计算以执行者为基础分布到多个服务器。

2. 动态资产加载

利用空间信息，Gazebo可以自动加载和卸载模拟资产，以显著提高性能。此功能与分布式仿真很好地配对。

3. 性能可调

控制模拟时间步长以实时、快于实时或慢于实时的方式运行。

4. 跨平台支持

可在Windows、Linux和MacOS上使用Gazebo库。

5. 云集成

在云托管服务器app.gazebosim.org上查看、下载和上传模拟模型和场景模型。

6. ROS集成

支持与 ROS Melodic通信，可以方便地进行数据转换。

7. 传感器和噪声模型

单目相机、深度相机、激光雷达、惯性传感器（IMU）、接触式传感器、高度计传感器和磁力计传感器均可用，更多传感器正在开发中。每个传感器可以选择性地利用噪声模型来注入高斯或定制噪声属性。

8. 高级3D图形

Gazebo支持OGRE 2.1渲染，它提供了访问最新渲染技术的途径，包括增强的阴影贴图、PBR材质和更快的渲染管道。

9. 精密物理

DART是Gazebo物理中的默认物理引擎，提供了超过游戏引擎的精确度。

二、Gazebo安装

本书使用Gazebo 11作为仿真环境。安装命令如下：

```
$ sudo apt install gazebo
```

如果要运行Gazebo，直接在终端中输入gazebo，如图1-2-19所示。

```
$ gazebo
```

Gazebo官方提供了大量模型可供使用，使用如下命令安装。

```
$ cd ~/.gazebo && wget https://gitee.com/clk_china/scripts/raw/master/gazebo_model.py && python3 gazebo_model.py
```

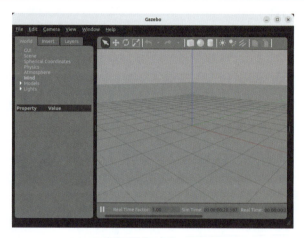

图1-2-19　Gazebo

三、Gazebo与ROS2集成

Gazebo是一个独立的软件，要与ROS2集成，还需要安装一些依赖。

1. 安装gazebo_ros_pkgs

gazebo_ros_pkgs是ROS2和Gazebo之间的桥梁，如图1-2-20所示，使用下面的指令一键安装。

```
$ sudo apt install ros-humble-gazebo-*
```

图1-2-20　Gazebo与ROS2集成原理

2. 使用ROS2控制Gazebo

首先启动Gazebo，并加载一个world文件，如图1-2-21所示。

```
$ gazebo /opt/ros/humble/share/gazebo_plugins/worlds/gazebo_ros_diff_drive_demo.world
```

Gazebo会在启动时直接加载指定的world文件，显示一个两轮小车。

使用ROS2命令驱动小车，打开一个新的终端，命令如下：

```
$ ros2 topic pub /demo/cmd_demo geometry_msgs/msg/Twist "{linear:
{x: 0.2,y: 0,z: 0},angular: {x: 0,y: 0,z: 0}}"
```

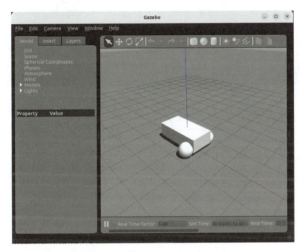

图1-2-21　Gazebo启动指定world文件

终端中输出如下信息：

```
publisher: beginning loop
publishing #1: geometry_msgs.msg.Twist(linear=geometry_msgs.msg.Vector3(x=0.2, y=0.0, z=0.0),
  angular=geometry_msgs.msg.Vector3(x=0.0, y=0.0, z=0.0))

publishing #2: geometry_msgs.msg.Twist(linear=geometry_msgs.msg.Vector3(x=0.2, y=0.0, z=0.0), angular=geometry_msgs.msg.Vector3(x=0.0, y=0.0, z=0.0))

publishing #3: geometry_msgs.msg.Twist(linear=geometry_msgs.msg.Vector3(x=0.2, y=0.0, z=0.0), angular=geometry_msgs.msg.Vector3(x=0.0, y=0.0, z=0.0))
...
```

可以看到Gazebo中的小车开始直线运行了。

思考与总结

1. Ubuntu系统中的常用命令有哪些？

2. ROS2是如何工作的？

3. ROS2可以通过专用工具与Gazebo集成起来，对模型的仿真和控制会非常方便。

视　频

仿真环境集成

项目三 ROS2入门

学习目标

1. 掌握ROS2工作空间与功能包的创建方式。
2. 掌握ROS2的工作方式。
3. 掌握ROS2的基本通信方式。

任务一 了解ROS2工作空间与常用命令

本任务目的是对ROS2的工作模式有一定的了解,学习使用ROS2工作空间和功能包。

一、工作空间

在使用一些开发环境,比如Visual Studio、Eclipse、Qt Creator、Keil等,在开始编写程序之前都会创建一个工程,此时就会产生一个文件夹,针对这个工程的所有文件都会放置在这个文件夹中,这个文件夹以及里边的内容,就称为工程,如图1-3-1所示。

在ROS2中进行开发时,各种代码文件、参数配置文件等也需要放在一个文件夹中进行管理,这个文件夹就是一个功能包。而工作空间就是包含若干个功能包的目录,ROS2依赖工作空间去查找功能包。

1. 创建工作空间

工作空间中必须包含一个src文件夹,用于存放ROS2功能包。

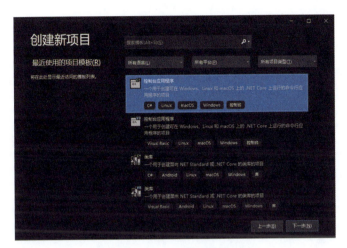

图1-3-1　Visual Studio创建项目

创建一个名为ros2_ws的工作空间，命令如下：

```
$ mkdir -p ~/ros2_ws/src
```

2. 创建功能包

功能包是包含某些特定功能的目录，ROS2可以直接执行。

ros2 pkg命令用于创建功能包。

```
$ ros2 pkg create <package> --build-type <type> --dependencies <dependency> --node-name <nodename>
```

命令中的参数见表1-3-1。

表1-3-1　ros2 pkg命令

命令	描述
create	创建功能包，<package>命名习惯是不出现大写字母
--build-type	编译的类型：cmake|ament_cmake|ament_python。Python选择ament_python
--dependencies	依赖，Python至少添加rclpy
--node-name	添加的节点名称，会自动创建节点文件，并在setup.py中声明该节点。只能设置一个节点，如果使用的节点多于一个，还需要手动添加

ROS2中功能包根据编译方式的不同分为三种类型：

① ament_python：适用于Python程序；

② cmake：适用于C++程序；

③ ament_cmake：适用于C++程序，是cmake的增强版。

ROS2中功能包根据语言种类的不同，主要有两种依赖：

① rclpy：适用于Python的依赖；

② rclcpp：适用于C++的依赖。

例如，创建一个名为pythondemo的功能包。

```
$ cd ~/ros2_ws/src
$ ros2 pkg create pythondemo --build-type ament_python --dependencies rclpy
```

创建完成后的pythondemo目录结构如下：

```
pythondemo
├── pythondemo
│   └── __init__.py
├── package.xml
├── resource
│   └── pythondemo
├── setup.cfg
├── setup.py
└── test
    ├── test_copyright.py
    ├── test_flake8.py
    └── test_pep257.py

3 directories, 8 files
```

3. 构建功能包

每当有功能包的修改时，都需要重新构建，方便工作空间能够识别到功能包。

ROS2使用colcon作为功能包构建工具，相当于ROS中的catkin工具。ROS2默认没有安装colcon，所以需要先安装：

```
$ sudo apt-get install python3-colcon-common-extensions
```

编译工作空间：

```
$ cd ~/ros2_ws
$ colcon build
```

自动将src下的所有软件包一次性构建：

```
Starting >>> pythondemo
--- stderr: pythondemo
/usr/lib/python3/dist-packages/setuptools/command/install.py:34:
SetuptoolsDeprecationWarning: setup.py install is deprecated. Use build and pip and other standards-based tools.
    warnings.warn(
---
Finished <<< pythondemo [0.67s]
```

```
Summary: 1 package finished [0.82s]
  1 package had stderr output: pythondemo
```

如果在构建中看到上述程序错误没关系,不影响使用,ROS2官方正在修复。错误原因是setuptools版本太高造成的。

构建完成后,在src同级目录会新增 build、install和log目录:

```
.
├── build
├── install
├── log
└── src

4 directories, 0 files
```

① build目录存储的是中间文件。对于每个包,将创建一个子文件夹,在其中调用例如cmake。

② install目录是每个软件包将安装到的位置。默认情况下,每个包都将安装到单独的子目录中。

③ log目录包含有关每个colcon调用的各种日志信息。

截至目前,仅是创建并构建了功能包,在本项目任务二和任务三中将学习添加节点、发布与订阅话题。

二、常用命令

在创建功能包的时候使用了ros2 pkg命令,一些其他常用的命令见表1-3-2～表1-3-5,可以在命令后添加--help查看详细用法,如ros2 pkg create--help。

表1-3-2 包命令

命令	描述
ros2 pkg create	创建功能包
ros2 pkg list	列出所有的功能包
ros2 pkg executables	列出功能包的可执行文件

表1-3-3 节点命令

命令	描述
ros2 node list	查看当前活动的节点列表
ros2 node info	查看节点详细信息,包括订阅、发布的消息,开启的服务和动作等
ros2 run	节点运行工具
ros2 launch	launch工具,可以一次启动多个节点

表1-3-4　话题命令

命令	描述
ros2 topic list	返回系统中当前活动的所有话题的列表
ros2 topic type	查看话题消息类型
ros2 topic info	查看话题信息，包括消息类型、订阅者数量、发布者数量等
ros2 topic echo	实时在控制台显示话题内容
ros2 topic pub	手动向话题发布消息

表1-3-5　接口命令

命令	描述
ros2 interface list	显示系统内所有的接口，包括消息（Messages）、服务（Services）、动作（Actions）
ros2 interface show	显示指定接口的详细内容

任务二　学习ROS2节点

本任务目的是对ROS2的节点有一定的概念，学习如何创建和使用ROS2节点。

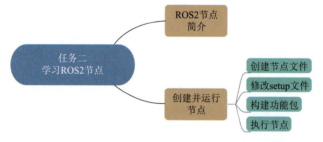

一、ROS2节点简介

ROS2节点就是ROS2功能包中用于实现特定功能的可执行文件，一个ROS2功能包可以包含多个ROS2节点。本书中，功能包均采用Python编写，所以节点就是一个py文件。

节点是ROS2中的最小单位，节点可以发布或接收一个话题，节点也可以提供或使用某种服务。

打开两个终端，分别启动海龟模拟器和控制节点，如图1-3-2所示。

启动节点使用的是ros2 run命令。

```
$ ros2 run turtlesim turtlesim_node
$ ros2 run turtlesim turtle_teleop_key
```

图1-3-2 海龟模拟器和控制节点

查看当前活动的节点列表：

$ ros2 node list

终端中输出如下信息：

/teleop_turtle
/turtlesim

查看节点信息：

$ ros2 node info /turtlesim

终端中输出如下信息：

/turtlesim
  Subscribers:
    /parameter_events: rcl_interfaces/msg/ParameterEvent
    /turtle1/cmd_vel: geometry_msgs/msg/Twist
  Publishers:
    /parameter_events: rcl_interfaces/msg/ParameterEvent
    /rosout: rcl_interfaces/msg/Log
    /turtle1/color_sensor: turtlesim/msg/Color
    /turtle1/pose: turtlesim/msg/Pose
  Service Servers:
    /clear: std_srvs/srv/Empty
    /kill: turtlesim/srv/Kill
    /reset: std_srvs/srv/Empty
    /spawn: turtlesim/srv/Spawn
    /turtle1/set_pen: turtlesim/srv/SetPen
    /turtle1/teleport_absolute: turtlesim/srv/TeleportAbsolute
    /turtle1/teleport_relative: turtlesim/srv/TeleportRelative

```
        /turtlesim/describe_parameters: rcl_interfaces/srv/DescribeParameters
        /turtlesim/get_parameter_types: rcl_interfaces/srv/GetParameterTypes
        /turtlesim/get_parameters: rcl_interfaces/srv/GetParameters
        /turtlesim/list_parameters: rcl_interfaces/srv/ListParameters
        /turtlesim/set_parameters: rcl_interfaces/srv/SetParameters
        /turtlesim/set_parameters_atomically: rcl_interfaces/srv/SetParametersAtomically
    Service Clients:

    Action Servers:
        /turtle1/rotate_absolute: turtlesim/action/RotateAbsolute
    Action Clients:
```

二、创建并运行节点

下面在本项目任务一创建的功能包中，创建自己的节点。

节点文件必须放在__init__.py同级目录中。

1. 创建节点文件

在__init__.py同级目录中，创建一个名为demo.py的文件，文件名可以自定义，但请注意不要包含中文和中文字符。

```python
#!/usr/bin/env python3
import rclpy
from rclpy.node import Node

class Demonode(Node):
    def __init__(self,name):
        super().__init__(name)
        self.count = 0
        self.get_logger().info("{}节点启动了".format(name))
        # 创建一个计时器回调，计时间隔0.5s
        self.timer = self.create_timer(0.5, self.timer_callback)

    def timer_callback(self):
        """计时器回调函数"""
        self.get_logger().info('这是第{}次回调'.format(self.count))
                            #打印一下发布的数据
        self.count += 1



def main(args=None):
    rclpy.init()                        # 初始化rclpy
    node = Demonode("demo")             # 新建一个节点
```

```
rclpy.spin(node)           # 保持节点运行，检测是否收到退出指令
                           （Ctrl+C）
rclpy.shutdown()           # 关闭rclpy
```

2. 修改setup文件

创建节点后，需要在setup.py中添加声明，这样在构建的时候才能找到该节点。

如下所示，添加'demo=pythondemo.demo:main'，表示在构建后生成一个可执行文件demo，指向pythondemo/demo.py中的main函数。

```
entry_points={
    'console_scripts': [
'demo = pythondemo.demo:main'
    ],
},
```

3. 构建功能包

修改节点后需要重新构建功能包。

```
$ cd ~/ros2_ws
$ colcon build
```

4. 执行节点

执行功能包前，需要source一下当前工作空间中的环境变量。环境变量仅对当前终端有效，新打开终端的话，需要重新source。

```
$ source install/setup.bash
$ ros2 run pythondemo demo
```

终端中输出如下信息：

```
[INFO] [1681981630.122434895] [demo]: demo节点启动了
[INFO] [1681981630.623920341] [demo]: 这是第0次回调
[INFO] [1681981631.123787430] [demo]: 这是第1次回调
[INFO] [1681981631.624489612] [demo]: 这是第2次回调
[INFO] [1681981632.123899354] [demo]: 这是第3次回调
[INFO] [1681981632.623877488] [demo]: 这是第4次回调
[INFO] [1681981633.123852965] [demo]: 这是第5次回调
[INFO] [1681981633.624441352] [demo]: 这是第6次回调
```

打印节点列表：

```
$ ros2 node list
```

终端中输出如下信息：

```
/demo
```

任务三　学习ROS2话题

本任务目的是对ROS2的话题有一定的了解，学习如何通过话题发布与订阅消息。

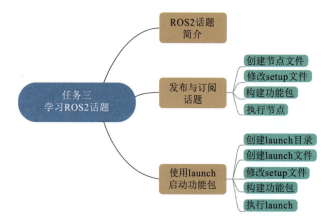

一、ROS2话题简介

话题是ROS2中最常用的通信方式，如图1-3-3所示，激光雷达、差速驱动和摄像头等都是通过话题来传递数据的。

一个节点发布数据到某个话题上，另外一个节点就可以通过订阅话题得到数据。

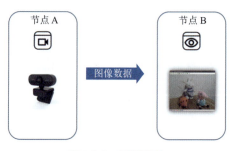

图1-3-3　话题通信

ROS2话题通信支持多种模式，可以是1对1，如图1-3-4所示；1对n，如图1-3-5所示；n对n，如图1-3-6所示；n对1，如图1-3-7所示；甚至是订阅自身发布的话题，如图1-3-8所示。

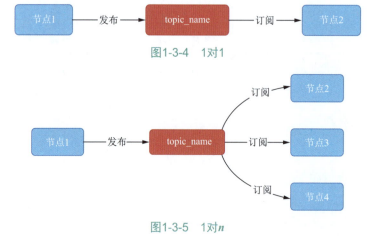

图1-3-4　1对1

图1-3-5　1对n

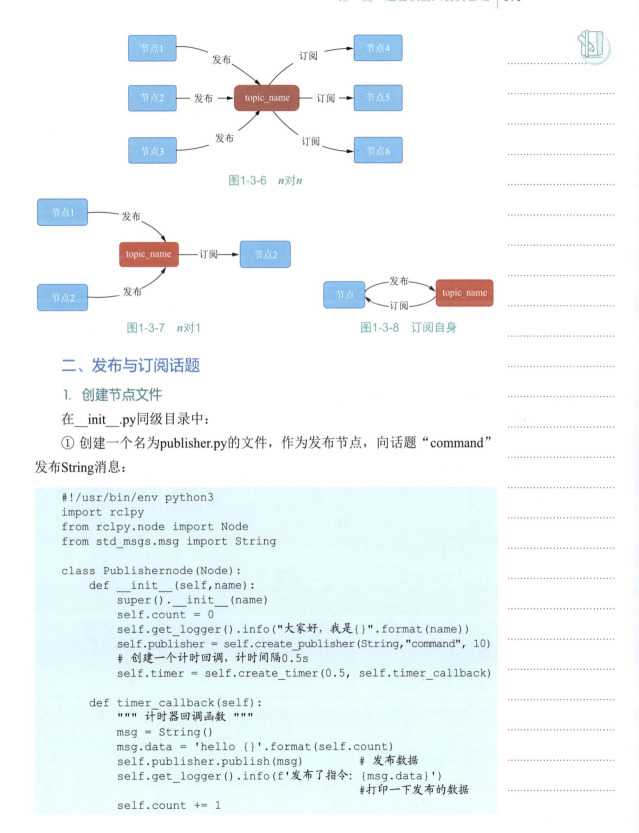

图1-3-6　n对n

图1-3-7　n对1

图1-3-8　订阅自身

二、发布与订阅话题

1. 创建节点文件

在__init__.py同级目录中：

① 创建一个名为publisher.py的文件，作为发布节点，向话题"command"发布String消息：

```python
#!/usr/bin/env python3
import rclpy
from rclpy.node import Node
from std_msgs.msg import String

class Publishernode(Node):
    def __init__(self,name):
        super().__init__(name)
        self.count = 0
        self.get_logger().info("大家好，我是{}".format(name))
        self.publisher = self.create_publisher(String,"command", 10)
        # 创建一个计时回调，计时间隔0.5s
        self.timer = self.create_timer(0.5, self.timer_callback)

    def timer_callback(self):
        """ 计时器回调函数 """
        msg = String()
        msg.data = 'hello {}'.format(self.count)
        self.publisher.publish(msg)           # 发布数据
        self.get_logger().info(f'发布了指令: {msg.data}')
                                              #打印一下发布的数据
        self.count += 1
```

```python
def main(args=None):
    rclpy.init(args=args)                              # 初始化rclpy
    node = Publishernode("publisher_node") # 新建一个节点
    rclpy.spin(node)        # 保持节点运行，检测是否收到退出指令（Ctrl+C）
    rclpy.shutdown()                                   # 关闭rclpy
```

② 创建一个名为subscriber.py的文件，作为订阅节点，订阅话题"command"，消息类型为String：

```python
#!/usr/bin/env python3
import rclpy
from rclpy.node import Node
from std_msgs.msg import String

class Subscribernode(Node):
    def __init__(self,name):
        super().__init__(name)
        self.get_logger().info("大家好，我是{}".format(name))
        # 订阅话题
        self.subscribe = self.create_subscription(String,"command",self.command_callback,10)

    def command_callback(self, msg):
        self.get_logger().info('收到消息,内容是{}'.format(msg.data))

def main(args=None):
    rclpy.init(args=args)                              # 初始化rclpy
    node = Subscribernode("subscriber_node")# 新建一个节点
    rclpy.spin(node)        # 保持节点运行，检测是否收到退出指令（Ctrl+C）
    rclpy.shutdown()                                   # 关闭rclpy
```

2. 修改setup文件

在setup.py中添加声明，使ROS2能够找到发布节点和订阅节点。

```python
    entry_points={
        'console_scripts': [
'publisher = pythondemo.publisher:main',
'subscriber = pythondemo.subscriber:main',
        ],
    },
```

生成一个可执行文件publisher，指向pythondemo/publisher.py中的main函数。
生成一个可执行文件subscriber，指向pythondemo/subscriber.py中的main函数。

3. 构建功能包

修改节点后需要重新构建功能包。

```
$ cd ~/ros2_ws
$ colcon build
```

4. 执行节点

使用source命令加载当前工作空间中的环境变量，执行订阅节点。

```
$ source install/setup.bash
$ ros2 run pythondemo subscriber
```

终端中输出如下信息：

```
[INFO] [1682039177.737676796] [subscriber_node]：大家好，我是 subscriber_node
```

不再输出新的内容了，这是因为还没有启动发布节点，订阅节点没有获取到数据。

新打开一个终端，执行发布节点，如图1-3-9所示。

```
$ cd ~/ros2_ws
$ source install/setup.bash
$ ros2 run pythondemo publisher
```

图1-3-9 话题通信

打印节点列表。

```
$ ros2 node list
```

终端中输出如下信息：

```
/publisher_node
/subscriber_node
```

打印话题列表。

```
$ ros2 topic list
```

终端中输出如下信息：

```
/command
/parameter_events
```

```
/rosout
```

订阅节点和发布节点开始同步更新，从而实现了话题通信。

三、使用launch启动功能包

在"二、发布与订阅话题"中，使用了两个终端，分别启动发布节点和订阅节点，实现话题通信。如果一个项目中包含大量节点，这种操作方式是非常不方便的。ROS2提供了另一种启动节点的方式，即ros2 launch命令，可以一次启动多个节点。

1. 创建launch目录

ros2 launch命令使用launch文件来一次启动多个节点。ROS2官方推荐将launch文件放在launch文件夹中，便于维护。

进入功能包目录，创建launch文件夹。

```
$ cd ~/ros2_ws/src/pythondemo
$ mkdir -p launch
```

2. 创建launch文件

launch文件实际就是一个py文件，官方推荐的命名方式是***.launch.py。

使用VScode在launch中创建名为pythondemo.launch.py的文件，编写以下代码：

```python
from launch import LaunchDescription
from launch_ros.actions import Node

def generate_launch_description():
    """
    1.固定的函数名，ROS2会对该函数名字做识别
    2.launch内容描述函数，由ros2 launch扫描调用
    """
    publisher = Node(
        package="pythondemo",
        executable="publisher"
        )
    subscriber = Node(
        package="pythondemo",
        executable="subscriber"
        )
    # 创建LaunchDescription对象launch_description，用于描述launch文件
    launch_description = LaunchDescription([publisher, subscriber])
    # 返回让ROS2根据launch描述执行节点
    return launch_description
```

3. 修改setup文件

launch文件相对于功能包来说是一个数据文件，必须在setup文件中添加对

它的安装，否则构建后将无法找到launch文件。修改后完整的setup文件如下：

```python
from setuptools import setup
from glob import glob
import os
package_name = 'pythondemo'

setup(
    name=package_name,
    version='0.0.0',
    packages=[package_name],
    data_files=[
        ('share/ament_index/resource_index/packages',
            ['resource/' + package_name]),
        ('share/' + package_name, ['package.xml']),
        (os.path.join('share', package_name, 'launch'), glob('launch/*py')),    # 添加对launch文件的安装
    ],
    install_requires=['setuptools'],
    zip_safe=True,
    maintainer='clk',
    maintainer_email='clk@todo.todo',
    description='TODO: Package description',
    license='TODO: License declaration',
    tests_require=['pytest'],
    entry_points={
        'console_scripts': [
    'publisher = pythondemo.publisher:main',
    'subscriber = pythondemo.subscriber:main',
        ],
    },
)
```

4. 构建功能包

重新构建功能包。

```
$ cd ~/ros2_ws
$ colcon build
```

5. 执行launch

加载环境变量，使用ros2 launch启动launch文件。

ros2 launch命令的详细用法，请通过ros2 launch --help查看。

```
$ source install/setup.bash
$ ros2 launch pythondemo pythondemo.launch.py
```

终端中输出如下信息：

```
[INFO] [launch]: All log files can be found below /home/clk/.ros/log/2023-04-21-09-30-19-670423-CLK-Ubuntu-6953
```

```
[INFO] [launch]: Default logging verbosity is set to INFO
[INFO] [publisher-1]: process started with pid [6954]
[INFO] [subscriber-2]: process started with pid [6956]
[subscriber-2] [INFO] [1682040619.879770345] [subscriber_node]: 大家好，我是subscriber_node
[publisher-1] [INFO] [1682040619.879963830] [publisher_node]: 大家好，我是publisher_node
[publisher-1] [INFO] [1682040620.382284307] [publisher_node]: 发布了指令: hello 0
[subscriber-2] [INFO] [1682040620.382986365] [subscriber_node]: 收到消息，内容是hello 0
[publisher-1] [INFO] [1682040620.882385035] [publisher_node]: 发布了指令: hello 1
[subscriber-2] [INFO] [1682040620.883157585] [subscriber_node]: 收到消息，内容是hello 1
[publisher-1] [INFO] [1682040621.381542844] [publisher_node]: 发布了指令: hello 2
[subscriber-2] [INFO] [1682040621.382035717] [subscriber_node]: 收到消息，内容是hello 2
[publisher-1] [INFO] [1682040621.881919999] [publisher_node]: 发布了指令: hello 3
[subscriber-2] [INFO] [1682040621.883140963] [subscriber_node]: 收到消息，内容是hello 3
[publisher-1] [INFO] [1682040622.382903308] [publisher_node]: 发布了指令: hello 4
[subscriber-2] [INFO] [1682040622.383772402] [subscriber_node]: 收到消息，内容是hello 4
[publisher-1] [INFO] [1682040622.882459446] [publisher_node]: 发布了指令: hello 5
[subscriber-2] [INFO] [1682040622.883335239] [subscriber_node]: 收到消息，内容是hello 5
…
```

可以看到一次性启动了两个节点，实现了话题通信。

打印节点列表。

```
$ ros2 node list
```

终端中输出如下信息：

```
/publisher_node
/subscriber_node
```

打印话题列表。

```
$ ros2 topic list
```

终端中输出如下信息：

```
/command
/parameter_events
/rosout
```

项目四

Gazebo入门

学习目标

1. 掌握Gazebo的基本操作方式。
2. 掌握搭建3D模型。

任务一　掌握Gazebo基本操作

本任务目的是对Gazebo有一定的了解，掌握基本的操作方法。

一、Gazebo功能介绍

1. Gazebo界面介绍与使用

在项目二任务三中已经介绍过Gazebo，现在直接启动：

```
$ gazebo
```

如图1-4-1所示，1是菜单栏，2是工具栏，3是左侧面板，4是仿真视图，其中的红色线、绿色线和蓝色线是Gazebo的坐标轴，分别对应x轴、y轴和z轴。

（1）菜单栏

主要使用File菜单和Edit菜单。

File菜单：包含保存和退出的功能，如图1-4-2所示。

Edit菜单：编辑菜单，如图1-4-3所示。

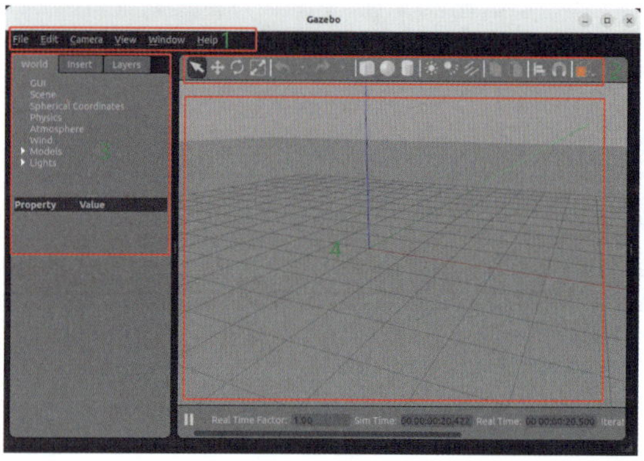

图1-4-1　Gazebo 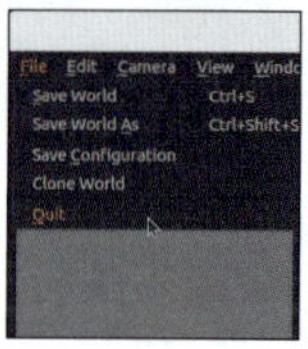　　　　　图1-4-2　File菜单 　图1-4-3　Edit菜单

（2）工具栏

包含一些和Gazebo交互时需要的功能，从左到右，见表1-4-1。

表1-4-1　工具栏对照表

序号	描述
1	选择模式：默认模式，在场景中做标注
2	移动模式：可以沿x轴、y轴、z轴，或任意方向移动模型
3	旋转模式：可以沿x轴、y轴、z轴旋转模型
4	缩放模式：可以沿x轴、y轴、z轴缩小或放大模型
5	撤销
6	重做
7	放置一个长方体
8	放置一个球体
9	放置一个圆柱体
10	放置点光源（球状点光源）
11	放置聚光灯（从上而下，金字塔状向下照射）
12	放置方向性光源（平行光）
13	复制
14	粘贴
15	对齐：将模型彼此对齐
16	捕捉：将一个模型捕捉到另一个模型
17	更改视图：从不同的角度看场景

（3）左侧面板

World：显示当前场景中的所有模型，可以通过这个选项查看和修改其参数。

Insert：向当前场景添加新的模型。

Layers：组织并显示仿真中可用的不同可视化组，很少使用。

（4）仿真视图

这是Gazebo的仿真窗口，所有的模型都在这里显示。

2. Gazebo系统命令

执行Gazebo帮助命令，查看帮助信息。

```
$ gazebo -h
```

Gazebo用法：gazebo [options] <world_file>。

Gazebo在启动的时候可以指定world文件，即在项目二任务三中显示的两轮小车。同时，Gazebo在启动时还可以添加一些配置信息，常用的配置见表1-4-2。

表1-4-2　常用的配置

参　　数	描　　述
-v [--version]	输出版本信息
--verbose	增加写入到终端的消息
-h [--help]	显示帮助信息
-u [--pause]	以暂停状态启动服务器
-e [--physics] arg	指定一个物理引擎，ode|bullet|dart|simbody
--gui-client-plugin arg	加载GUI插件
-s [--server-plugin] arg	加载服务器插件

在启动ROS2插件的时候，就需要使用-s参数添加启动ROS2插件。命令如下：

```
$ gazebo --verbose -s libgazebo_ros_init.so -s libgazebo_ros_factory.so
```

libgazebo_ros_init.so和libgazebo_ros_factory.so就是ROS2插件，启动后才可以使用ROS2命令对Gazebo进行操作。

二、Gazebo绘制操作

1. 绘制模型

（1）添加模型

使用鼠标左键单击工具栏上的图标，然后在仿真视图中的合适位置单击，放置模型，如图1-4-4所示。

视　频

仿真模型绘制

图1-4-4 添加模型

单击工具栏中的移动、旋转或缩放按钮，如图1-4-5所示，就可以进入对应的模型，对模型的位置、方向和大小进行操作。

图1-4-5 变换模型

（2）编辑模型

回到选择模式，在模型上右击，选择Edit model命令，如图1-4-6所示，进入模型编辑视图。

在模型编辑视图中双击模型，如图1-4-7所示，即可打开模型编辑对话框，可以对Link、Visual和Collision进行详细编辑。

单击File菜单，保存并退出模型编辑。

2. 绘制建筑

在Edit菜单中，选择Building Editor，进入建筑编辑。Gazebo默认提供Wall、Window、Door和Stairs。下面使用Wall绘制一个简单墙壁。

图1-4-6　选择Edit model命令

图1-4-7　模型编辑

鼠标左键单击Wall图标，在视图上方的白色绘制区域，单击左键开始绘制，再次单击左键结束绘制，重复操作，绘制多个墙壁，单击右键退出绘制，如图1-4-8所示。

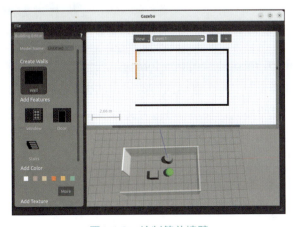

图1-4-8　绘制简单墙壁

单击File菜单，保存并退出建筑编辑。

再次单击File菜单，选择Save World As命令，将建筑和模型一起保存到Gazebo的world文件中，例如命名为demo.world，如图1-4-9所示。

```
$ gazebo demo.world
```

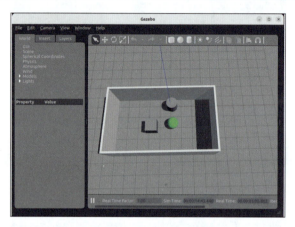

图1-4-9　打开指定world

任务二　学习URDF模型搭建

本任务目的是对URDF文件有一定的了解，学习编写一个简单的机器人URDF文件。

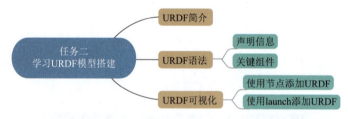

一、URDF简介

URDF（unified robot description format，统一机器人描述格式）与计算机文件中的.txt文本格式、.jpg图像格式等类似，URDF是一种基于XML规范、用于描述机器人结构的格式，使用urdf作为文件后缀。设计这一格式的目的在于提供一种尽可能通用的机器人描述规范。

ROS官网网址，请自行在搜索引擎中搜索获得。

一个简单的示例，demo.urdf，如下所示：

```
<?xml version="1.0"?>
```

```
<robot name="mybot">
    <link name="base_link">
        <visual>
            <geometry>
                <cylinder length="0.6" radius="0.2"/>
            </geometry>
        </visual>
    </link>
</robot>
```

从机构学角度讲，机器人通常被建模为由连杆（link）和关节（joint）组成的结构。连杆是带有质量属性的刚体，而关节是连接、限制两个刚体相对运动的结构。通过关节将连杆依次连接起来，就构成了一个个运动链（也就是这里所定义的机器人模型）。一个URDF文档即描述了这样的一系列关节与连杆的相对关系、惯性属性、几何特点和碰撞模型。具体来说，包括：

① 机器人模型的运动学与动力学描述。

② 机器人的几何表示。

③ 机器人的碰撞模型。

二、URDF语法

一个完整的URDF，由声明信息和关键组件共同组成：

```
<?xml version="1.0"?>
<robot name="myrobot">
    <link>…</link>
    <link>…</link>
    …
    <joint>…</joint>
    <joint>…</joint>
</robot>
```

对于标签的详细使用方法请查看官网。

1. **声明信息**

两种声明信息分别是XML声明和机器人声明。

（1）XML声明

URDF是一种基于XML规范的描述语言，因此需要声明这是一个xml=文件，表明其版本号，一般位于URDF文件第一行。

```
<?xml version="1.0"?>
```

（2）机器人声明

声明机器人的所有内容。

robot标签声明一个机器人模型，标签对内部包含了所有的组件信息。

```
<robot name="myrobot">
    ...
</robot>
```

2. 关键组件

主要有两种组件：link组件和joint组件。

（1）link组件

link组件即连杆，机器人的每一部分就是一个link，例如一个简单的两轮小车，躯干、左轮、右轮、支撑轮，都是一个link，如图1-4-10所示。

一个link元素必须包含visual、collision和inertial三部分，一个简单的示例代码如下：

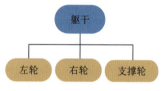

图1-4-10　机器人link示意图

```
<!-- base link -->
<link name="base_link">
    <visual>
        <origin xyz="0 0 0.0" rpy="0 0 0" />
        <geometry>
            <cylinder length="0.12" radius="0.10" />
        </geometry>
        <material name="blue">
            <color rgba="0.1 0.1 1.0 0.5" />
        </material>
    </visual>
    <collision>
        <origin xyz="0 0 0.0" rpy="0 0 0" />
        <geometry>
            <cylinder length="0.12" radius="0.10" />
        </geometry>
        <material name="blue">
            <color rgba="0.1 0.1 1.0 0.5" />
        </material>
    </collision>
    <inertial>
        <mass value="0.2" />
        <inertia ixx="0.0122666" ixy="0" ixz="0" iyy="0.0122666" iyz="0" izz="0.02" />
    </inertial>
</link>
```

（2）joint组件

机器人的link与link之间可能存在连接关系，这个连接称为joint（关节）。对于上述的两轮机器人，左轮、右轮和支撑轮都是通过joint与躯干连接起来的，如图1-4-11所示。

图1-4-11　joint示意图

在joint中需要指明父link和子link，并确定两者的连接方式，一个简单的示例代码如下：

```
<joint name="left_wheel_joint" type="continuous">
    <parent link="base_link" />
    <child link="left_wheel_link" />
    <origin xyz="-0.02 0.10 -0.06" />
    <axis xyz="0 1 0" />
</joint>
```

三、URDF可视化

在使用Gazebo仿真时，使用world文件来加载静态资源，例如房屋、道路、迷宫等；使用URDF文件来加载动态的模型，如机器人。

Gazebo在启动时可以直接指定加载的world文件，而URDF文件无法直接加载，需要通过ROS2内置的功能包来添加到Gazebo中。

完整版URDF文件demo.urdf，请扫描二维码获取。

1. 使用节点添加URDF

启动Gazebo和ROS2插件。

```
$ gazebo --verbose -s libgazebo_ros_init.so -s libgazebo_ros_factory.so demo.world
```

若没有指定world文件，gazebo会自动使用安装目录的empty.world（一个空的world文件）。

新打开一个终端，启动ROS2中用于添加URDF的功能包，如图1-4-12所示。

```
$ ros2 run gazebo_ros spawn_entity.py -entity TQD_Maze robot  -file demo.urdf
```

gazebo_ros：用于添加URDF的功能包。

spawn_entity.py：节点文件。

-entity TQD_Maze robot：指定机器人的名字。

-file demo.urdf：指定urdf文件地址。

这时就可以在Gazebo中看到机器人了。

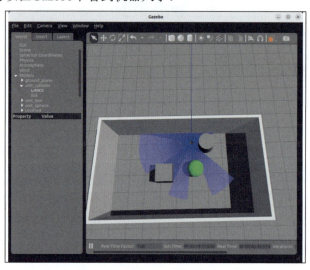

图1-4-12　使用节点添加URDF

2. 使用launch添加URDF

在话题通信中已经介绍过，launch可以一次性启动多个节点。

启动Gazebo加载world，添加URDF文件，均可以按照启动一个节点来对待，所以可以通过launch直接完成。

在pythondemo功能包中完成以下操作：

新建world目录，用于存放world文件，将demo.world复制到world文件夹中；

新建urdf目录，用于存放urdf文件，将demo.urdf复制到urdf文件夹中。

（1）修改launch文件

添加启动gazebo，并直接加载demo.world；

添加spawn_entity.py节点，添加demo.urdf。

```
import os
from launch import LaunchDescription
from launch.substitutions import LaunchConfiguration
from launch_ros.actions import Node
from launch.actions import ExecuteProcess
from launch_ros.substitutions import FindPackageShare

def generate_launch_description():
    """
    1.固定的函数名，ROS2会对该函数名字做识别
    2.launch内容描述函数，由ros2 launch 扫描调用
    """
```

```python
        robot_name_in_model = 'TQD_Robot'
        package_name = 'pythondemo'
        world_name = 'demo.world'
        urdf_name = "demo.urdf"
        pkg_share = FindPackageShare(package=package_name).find(package_name)
        world_model_path = os.path.join(pkg_share, f'world/{world_name}')
        urdf_model_path = os.path.join(pkg_share, f'urdf/{urdf_name}')

        # 启动Gazebo，加载world
        start_gazebo_cmd = ExecuteProcess(
            cmd=['gazebo', '--verbose', '-s', 'libgazebo_ros_init.so', '-s', 'libgazebo_ros_factory.so', world_model_path],
            output='screen')

        # 添加URDF
        spawn_entity_cmd = Node(
            package='gazebo_ros',
            executable='spawn_entity.py',
            arguments=['-entity', robot_name_in_model, '-file', urdf_model_path], output='screen')
        ld = LaunchDescription()
        ld.add_action(start_gazebo_cmd)
        ld.add_action(spawn_entity_cmd)
        return ld
```

（2）修改setup文件

添加对world文件的安装。

添加对urdf文件的安装。

```python
from setuptools import setup
from glob import glob
import os
package_name = 'pythondemo'

setup(
    name=package_name,
    version='0.0.0',
    packages=[package_name],
    data_files=[
        ('share/ament_index/resource_index/packages',
            ['resource/' + package_name]),
        ('share/' + package_name, ['package.xml']),
        (os.path.join('share', package_name, 'launch'), glob('launch/**')),
# 安装launch文件
        (os.path.join('share', package_name, 'world'), glob('world/**')),
# 安装world文件
        (os.path.join('share', package_name, 'urdf'), glob('urdf/**'))
# 安装urdf文件
    ],
```

```
install_requires=['setuptools'],
zip_safe=True,
maintainer='clk',
maintainer_email='clk@todo.todo',
description='TODO: Package description',
license='TODO: License declaration',
tests_require=['pytest'],
entry_points={
    'console_scripts': [
'publisher = pythondemo.publisher:main',
'subscriber = pythondemo.subscriber:main',
'demo = pythondemo.demo:main'
    ],
},
)
```

（3）构建启动

```
$ cd ~/ros2_ws
$ colcon build
$ source install/setup.bash
$ ros2 launch pythondemo pythondemo.launch.py
```

从而实现了使用launch直接加载world和urdf，如图1-4-13所示。

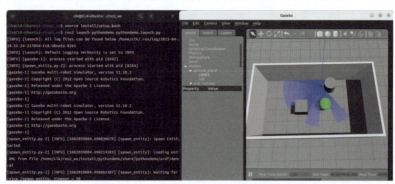

图1-4-13　使用launch添加URDF

思考与总结

1. 如何通过程序控制Gazebo中的3D模型？
2. Gazebo通过命令行参数可以非常方便地加载指定的地图。

第二篇 迷宫机器人仿真设计

进行迷宫机器人仿真，需要设计一个迷宫机器人模型，然后使用软件进行3D渲染，最后使用程序进行控制。本篇主要讲解制作一个迷宫机器人模型，在Gazebo中渲染显示出来；通过ROS通信，控制模型实现各种动作。

项目一

学习迷宫机器人制作

学习目标

1. 掌握制作国际标准的虚拟仿真迷宫机器人的方法。
2. 掌握制作国际标准的虚拟仿真迷宫的方法。

任务一　制作虚拟仿真迷宫机器人

本任务目的是学习制作自己的虚拟仿真迷宫机器人。

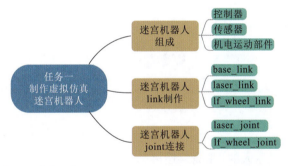

一、迷宫机器人组成

现实当中的迷宫机器人由三大部分组成，控制器、传感器和机电运动部件。真实迷宫机器人如图2-1-1所示。基本功能是从起点运行到终点。传感器在运行时检测四周墙壁信息，控制器对这些数据进行处理，结合自身算法，控制机电运动部件运行。

图2-1-1　真实迷宫机器人

虚拟仿真迷宫机器人是由ROS2和Gazebo根据URDF模型文件仿真出来的虚拟的迷宫机器人，同样包含这三部分。

1. 控制器

ROS2话题和消息、ROS2预留话题通信API，可以实现对虚拟仿真迷宫机器人的控制。

2. 传感器

Gazebo插件、激光雷达插件，可以实现红外线传感器的功能，检测周围障碍物的距离。

3. 机电运动部件

Gazebo插件、差速驱动插件，可以实现电机的功能，驱动模型直行和转弯。

因此，虚拟仿真迷宫机器人需要ROS2和Gazebo配合才能正常运行。

在后续内容中，若没有特殊说明，迷宫机器人均指代虚拟仿真迷宫机器人。

一个简单的四轮迷宫机器人示例如图2-1-2所示。

图2-1-2　四轮迷宫机器人示例

迷宫机器人共包含六个link，如图2-1-3所示。

图2-1-3　六个link

躯干是迷宫机器人的主link，连接有五个子link，激光link用于实现Gazebo中的激光雷达插件，四个轮子是迷宫机器人的运动结构。子link与主link，通过joint连接，组成一个完整的迷宫机器人。

二、迷宫机器人link制作

参考国际比赛规则，迷宫机器人在迷宫中运行，有尺寸的限制，过大或过小，都会对运行产生不便的影响。

迷宫单元格尺寸为18 cm×18 cm，挡板尺寸为18 cm×1.2 cm×5 cm，因此推荐的迷宫机器人尺寸为长9 cm，宽8 cm，高3 cm。

迷宫机器人link制作参考程序：

1. base_link

```
<link name="base_link">
```

```xml
<visual>    <!-- 视觉元素，描述link的形状、大小和颜色等 -->
    <origin rpy="0 0 0" xyz="0 0 0" />
    <geometry>
        <box size="0.09 0.06 0.01" />
    </geometry>
    <material name="blue">
        <color rgba="0.0 0.0 0.8 1.0" />
    </material>
</visual>
<collision>    <!-- 碰撞元素，描述与其他模型产生碰撞的参考 -->
    <origin rpy="0 0 0" xyz="0 0 0" />
    <geometry>
        <box size="0.09 0.06 0.01" />
    </geometry>
    <material name="green">
        <color rgba="0.0 0.8 0.0 1.0" />
    </material>
</collision>
<inertial>    <!-- 惯性元素，描述link的质量和转动惯性矩阵等 -->
    <origin rpy="1.5707963267948966 0 1.5707963267948966" xyz="0 0 0" />
    <mass value="0.4" />
    <inertia ixx="0.0001233333333333333" ixy="0.0" ixz="0.0" iyy="0.00038999999999999994"
             iyz="0.0" izz="0.0002733333333333333" />
</inertial>
</link>
```

2. laser_link

```xml
<link name="laser_link">
    <visual>    <!-- 视觉元素，描述link的形状、大小和颜色等 -->
        <origin rpy="0 0 0" xyz="0 0 0" />
        <geometry>
            <cylinder length="0.01" radius="0.005" />
        </geometry>
        <material name="blue">
            <color rgba="0.0 0.0 0.8 1.0" />
        </material>
    </visual>
    <collision>    <!-- 碰撞元素，描述与其他模型产生碰撞的参考 -->
        <origin rpy="0 0 0" xyz="0 0 0" />
        <geometry>
            <cylinder length="0.01" radius="0.005" />
        </geometry>
        <material name="blue">
            <color rgba="0.0 0.0 0.8 1.0" />
```

```xml
            </material>
        </collision>
        <inertial>   <!-- 惯性元素，描述link的质量和转动惯性矩阵等 -->
            <origin rpy="1.5707963267948966 0 0" xyz="0 0 0" />
            <mass value="0.1" />
            <inertia ixx="1.4583333333333333e-06" ixy="0" ixz="0" iyy="1.4583333333333333e-06"
                     iyz="0" izz="1.25e-06" />
        </inertial>
    </link>
```

3. lf_wheel_link

四个轮子link完全相同，注意命名为不同的name，例如rf_wheel_link、lr_wheel_link和rr_wheel_link。

```xml
    <link name="lf_wheel_link">
        <visual>     <!-- 视觉元素，描述link的形状、大小和颜色等 -->
            <origin rpy="1.57079 0 0" xyz="0 0 0" />
            <geometry>
                <cylinder length="0.01" radius="0.01" />
            </geometry>
            <material name="blue">
                <color rgba="0.0 0.0 0.8 1.0" />
            </material>
        </visual>
        <collision>    <!-- 碰撞元素，描述与其他模型产生碰撞的参考 -->
            <origin rpy="1.57079 0 0" xyz="0 0 0" />
            <geometry>
                <cylinder length="0.01" radius="0.01" />
            </geometry>
            <material name="blue">
                <color rgba="0.0 0.0 0.8 1.0" />
            </material>
        </collision>
        <inertial>   <!-- 惯性元素，描述link的质量和转动惯性矩阵等 -->
            <origin rpy="1.5707963267948966 0 0" xyz="0 0 0" />
            <mass value="0.1" />
            <inertia ixx="3.333333333333333e-06" ixy="0" ixz="0" iyy="3.333333333333333e-06" iyz="0"
                     izz="5e-06" />
        </inertial>
    </link>
```

link描述了迷宫机器人的外观、碰撞和惯性属性，如果参数不匹配或产生冲突，将对迷宫机器人的运动产生巨大的影响，因此，推荐使用TQD评测系统，一键制作，快速生成迷宫机器人link。

三、迷宫机器人joint连接

子link与主link的连接位置和连接方式，决定了该joint的类型。

迷宫机器人joint连接参考程序：

1. laser_joint

```
<joint name="laser_joint" type="fixed"> <!-- 激光link不需要转动，fixed -->
        <parent link="base_link" />
        <child link="laser_link" />
        <origin xyz="0.025 0 0.01" />
</joint>
```

2. lf_wheel_joint

四个轮子与主link的joint基本相同，唯一需要注意的是连接的位置是不同的。

```
<joint name="lf_wheel_joint" type="continuous">    <!-- 轮子link是无限旋转的，continuous -->
        <!-- left front-->
        <parent link="base_link" />
        <child link="lf_wheel_link" />
        <origin xyz="0.005 0.035 0" />
        <axis xyz="0 1 0" />
</joint>
```

完成link与joint后，一个迷宫机器人就做好了，现在它还没有传感器（激光雷达）和机电运动部件（差速驱动），这部分将在本篇项目二中进行介绍。

完整的迷宫机器人URDF文件，可以扫描二维码获取。

素 材

迷宫机器人URDF下载

任务二　制作虚拟仿真迷宫

本任务目的是学习制作虚拟仿真迷宫。

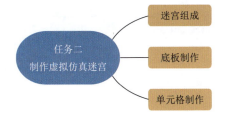

一、迷宫组成

一个符合IEEE国际标准的真实迷宫如图2-1-4所示。

共包含16×16个单元格，每个单元格尺寸为18 cm×18 cm，每一面墙壁的改变，都会组成一个全新的迷宫。

虚拟仿真迷宫与真实迷宫一样，同样包含底板和挡板，由于不需要手动插拔挡板，因此在虚拟仿真迷宫中将立柱合并到挡板中，如图2-1-5所示。

图2-1-4　真实迷宫

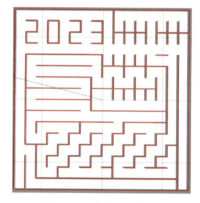
图2-1-5　虚拟迷宫示例

虚拟仿真迷宫底板是一个整体，直接按照长方体对待即可。单元格墙壁是独立的，推荐将迷宫第一行上方的墙壁和最后一列右侧的墙壁全部初始化绘制，对于每个单元格仅需考虑其左侧和下方的墙壁。

二、底板制作

底板是一个link，需要包含visual、collision和inertial三部分。

```xml
<visual name="visual">
    <pose frame="">0 0 0.0025 0 -0 0</pose>
    <geometry>
        <box>
            <size>2.892 2.892 0.005</size>
        </box>
    </geometry>
    <material>
        <script>
            <uri>file://media/materials/scripts/gazebo.material</uri>
            <name>Gazebo/White</name>
        </script>
    </material>
</visual>
<collision name="collision">
    <pose frame="">0 0 0.0025 0 -0 0</pose>
    <geometry>
        <box>
            <size>2.892 2.892 0.005</size>
```

```xml
            </box>
        </geometry>
        <surface>
            <friction>
                <ode>
                    <mu>1000</mu>
                    <mu2>1000</mu2>
                </ode>
                <torsional>
                    <ode />
                </torsional>
            </friction>
            <contact>
                <ode />
            </contact>
            <bounce />
        </surface>
        <max_contacts>10</max_contacts>
</collision>
<inertial>
    <mass>1</mass>
    <inertia>
        <ixx>1.39394</ixx>
        <ixy>0</ixy>
        <ixz>0</ixz>
        <iyy>0.696972</iyy>
        <iyz>0</iyz>
        <izz>0.696973</izz>
    </inertia>
    <pose frame="">0 0 0.05 0 -0 0</pose>
</inertial>
```

三、单元格制作

每个单元格包含四块挡板，迷宫第一行上方的墙壁和最后一列右侧的墙壁全部初始化绘制后，对于每个单元格仅需考虑其左侧和下方的墙壁。

绘制北面墙壁：

```xml
<visual name="paint_v_0_0_N">
    <pose frame="">-1.35 1.44 0.05 0 -0 0</pose>
    <geometry>
        <box>
            <size>0.192 0.012 0.01</size>
        </box>
    </geometry>
    <material>
        <script>
            <name>Gazebo/Red</name>
```

```xml
                <uri>file://media/materials/scripts/gazebo.material</uri>
            </script>
        </material>
    </visual>
    <visual name="v_0_0_N">
        <pose frame="">-1.35 1.44 0.025 0 -0 0</pose>
        <geometry>
            <box>
                <size>0.192 0.012 0.04</size>
            </box>
        </geometry>
    </visual>
    <collision name="p_0_0_N">
        <pose frame="">-1.35 1.44 0.03 0 -0 0</pose>
        <geometry>
            <box>
                <size>0.192 0.012 0.05</size>
            </box>
        </geometry>
        <max_contacts>10</max_contacts>
        <surface>
            <contact>
                <ode />
            </contact>
            <bounce />
            <friction>
                <torsional>
                    <ode />
                </torsional>
                <ode />
            </friction>
        </surface>
    </collision>
```

绘制东面墙壁：

```xml
    <visual name="paint_v_0_15_E">
        <pose frame="">1.44 1.35 0.05 0 -0 1.5708</pose>
        <geometry>
            <box>
                <size>0.192 0.012 0.01</size>
            </box>
        </geometry>
        <material>
            <script>
                <name>Gazebo/Red</name>
                <uri>file://media/materials/scripts/gazebo.material</uri>
```

```xml
        </script>
    </material>
</visual>
<visual name="v_0_15_E">
    <pose frame="">1.44 1.35 0.025 0 -0 1.5708</pose>
    <geometry>
        <box>
            <size>0.192 0.012 0.04</size>
        </box>
    </geometry>
</visual>
<collision name="p_0_15_E">
    <pose frame="">1.44 1.35 0.03 0 -0 1.5708</pose>
    <geometry>
        <box>
            <size>0.192 0.012 0.05</size>
        </box>
    </geometry>
    <max_contacts>10</max_contacts>
    <surface>
        <contact>
            <ode />
        </contact>
        <bounce />
        <friction>
            <torsional>
                <ode />
            </torsional>
            <ode />
        </friction>
    </surface>
</collision>
```

绘制西面墙壁：

```xml
<visual name="paint_v_0_0_W">
    <pose frame="">-1.44 1.35 0.05 0 -0 1.5708</pose>
    <geometry>
        <box>
            <size>0.192 0.012 0.01</size>
        </box>
    </geometry>
    <material>
        <script>
            <name>Gazebo/Red</name>
            <uri>file://media/materials/scripts/gazebo.material</uri>
        </script>
    </material>
```

```xml
        </visual>
        <visual name="v_0_0_W">
            <pose frame="">-1.44 1.35 0.025 0 -0 1.5708</pose>
            <geometry>
                <box>
                    <size>0.192 0.012 0.04</size>
                </box>
            </geometry>
        </visual>
        <collision name="p_0_0_W">
            <pose frame="">-1.44 1.35 0.03 0 -0 1.5708</pose>
            <geometry>
                <box>
                    <size>0.192 0.012 0.05</size>
                </box>
            </geometry>
            <max_contacts>10</max_contacts>
            <surface>
                <contact>
                    <ode></ode>
                </contact>
                <bounce></bounce>
                <friction>
                    <torsional>
                        <ode></ode>
                    </torsional>
                    <ode></ode>
                </friction>
            </surface>
        </collision>
```

绘制南面墙壁：

```xml
Draw the southern wall.<visual name="paint_v_0_1_S">
            <pose frame="">-1.17 1.26 0.05 0 -0 0</pose>
            <geometry>
                <box>
                    <size>0.192 0.012 0.01</size>
                </box>
            </geometry>
            <material>
                <script>
                    <name>Gazebo/Red</name>
                    <uri>file://media/materials/scripts/gazebo.material</uri>
                </script>
            </material>
        </visual>
        <visual name="v_0_1_S">
```

```xml
            <pose frame="">-1.17 1.26 0.025 0 -0 0</pose>
            <geometry>
                <box>
                    <size>0.192 0.012 0.04</size>
                </box>
            </geometry>
        </visual>
        <collision name="p_0_1_S">
            <pose frame="">-1.17 1.26 0.03 0 -0 0</pose>
            <geometry>
                <box>
                    <size>0.192 0.012 0.05</size>
                </box>
            </geometry>
            <max_contacts>10</max_contacts>
            <surface>
                <contact>
                    <ode></ode>
                </contact>
                <bounce></bounce>
                <friction>
                    <torsional>
                        <ode></ode>
                    </torsional>
                    <ode></ode>
                </friction>
            </surface>
        </collision>
```

由于迷宫共计256个单元格，手动编写代码难度较高，因此，推荐使用TQD评测系统，根据TQD模型文件一键制作，快速生成迷宫。完成迷宫的制作后，就可以在Gazebo中将其显示出来了。

完整的迷宫文件，可以扫描二维码获取。

素 材

迷宫文件下载

任务三　学习虚拟仿真系统启动

本任务目的是使用ROS2功能包启动自己制作的world文件和urdf文件，为以后进行迷宫机器人的开发奠定基础。

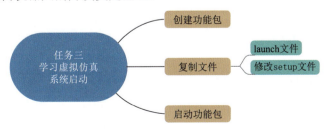

一、创建功能包

创建一个名为Maze robot _ws的工作空间，在其中创建一个名为tqd_Maze robot 的功能包，作为后续迷宫机器人开发的目录。

```
$ mkdir -p ~/Maze robot _ws/src
$ cd ~/Maze robot _ws/src
$ ros2 pkg create tqd_Maze robot  --build-type ament_python --dependencies rclpy
```

创建完成后的tqd_Maze robot 目录结构：

```
tqd_Maze robot
├── tqd_Maze robot
│   └── __init__.py
├── package.xml
├── resource
│   └── tqd_Maze robot
├── setup.cfg
├── setup.py
└── test
    ├── test_copyright.py
    ├── test_flake8.py
    └── test_pep257.py

3 directories, 8 files
```

二、复制文件

在tqd_Maze robot 功能包中完成以下操作：

新建launch目录，用于存放launch文件，新建tqd.launch.py作为启动文件。

新建urdf目录，用于存放urdf文件，将本项目任务一中制作的demo.urdf复制到urdf文件夹中。

新建world目录，用于存放world文件，将本项目任务二中制作的demo.world复制到world文件夹中。

1. launch文件

```python
from launch import LaunchDescription
from launch_ros.actions import Node
from launch.actions import ExecuteProcess
from launch_ros.substitutions import FindPackageShare
import os

def generate_launch_description():
    robot_name_in_model = 'TQD_Robot'
    package_name = 'tqd_Maze robot'
```

```
        urdf_name = 'demo.urdf'
        world_name = 'demo.world'

        ld = LaunchDescription()
        pkg_share = FindPackageShare(package=package_name).find(package_
name)
        urdf_model_path = os.path.join(pkg_share, f'urdf/{urdf_name}')
        world_model_path = os.path.join(pkg_share,f'world/{world_name}')

        # 启动Gazebo服务器
        start_gazebo_cmd = ExecuteProcess(
            cmd=['gazebo', '--verbose', '-s', 'libgazebo_ros_factory.
so', world_model_path],  # , world_model_path
            output='screen'
        )

        # 启动机器人
        spawn_entity_cmd = Node(
            package='gazebo_ros',
            executable='spawn_entity.py',
            arguments=['-entity', robot_name_in_model,  '-file', urdf_
model_path, '-x', '-1.35', '-y', '-1.35', '-z', '0.10', '-Y', '1.575'],
            output='screen'
        )

        ld = LaunchDescription()
        ld.add_action(start_gazebo_cmd)
        ld.add_action(spawn_entity_cmd)
        return ld
```

2. 修改setup文件

```
from setuptools import setup
from glob import glob
import os

package_name = 'tqd_Maze_robot'

setup(
    name=package_name,
    version='0.0.0',
    packages=[package_name],
    data_files=[
        ('share/ament_index/resource_index/packages',
            ['resource/' + package_name]),
        ('share/' + package_name, ['package.xml']),
        (os.path.join('share', package_name, 'launch'), glob('launch/
*.launch.py')),
        (os.path.join('share', package_name, 'urdf'), glob('urdf/
**')),
        (os.path.join('share', package_name, 'world'), glob('world/
**')),
```

```
    ],
    install_requires=['setuptools'],
    zip_safe=True,
    maintainer='clk',
    maintainer_email='clk@todo.todo',
    description='TODO: Package description',
    license='TODO: License declaration',
    tests_require=['pytest'],
    entry_points={
        'console_scripts': [
        ],
    },
)
```

三、启动功能包

```
$ cd ~/Maze_robot_ws
$ colcon build
$ source install/setup.bash
$ ros2 launch tqd_Maze_robot tqd.launch.py
```

启动 world 和 urdf 如图2-1-6所示。

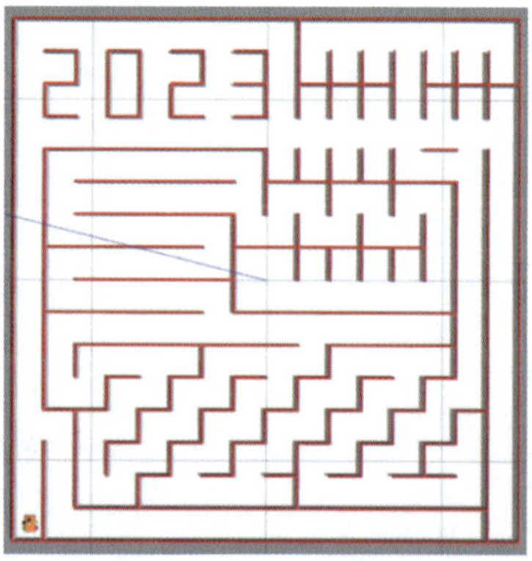

图2-1-6　启动world和urdf

思考与总结

1. 在制作迷宫机器人的时候，四个轮子应当如何摆放？

2. 在绘制迷宫的时候如何对每个单元格进行精准定位？

3. world和urdf中的迷宫是两个不同的模型，应当注意显示时两者的相对位置。

项目二 了解迷宫机器人基础功能

学习目标

1. 掌握两轮差速插件的原理与使用方式。
2. 掌握激光雷达插件的原理与使用方式。
3. 掌握多话题协同工作程序控制。

任务一 了解两轮差速驱动

本任务目的是学习使用两轮差速插件，驱动迷宫机器人运行。

一、两轮差速插件简介

机器人的驱动方式主要有电机驱动方式、液压驱动方式、气动驱动方式，如图2-2-1所示。

电机驱动器是利用各种电动机产生的力或转矩直接驱动机器人的关节，或者通过诸如减速机构来驱动机器人的关节，以获得所需的位置、速度、加速度和其他指标。具有环保、整洁、控制方便、运动精度高、维护成本低、驱动效率高的优点。

液压驱动器使用液体作为介质来传递力，并使用液压泵使液压系统产生的压力驱动执行器运动。液压驱动模式是成熟的驱动模式。

(a) 电机驱动方式 　　　　　(b) 液压驱动方式 　　　　　(c) 气动驱动方式

图2-2-1　机器人的三种驱动方式

气动驱动器使用空气作为工作介质，并使用气源发生器将压缩空气的压力能转换为机械能，以驱动执行器完成预定的运动。气动驱动具有节能、简单、时间短、动作快、柔软、质量小、产量/质量比高、安装维护方便、安全、成本低、对环境无污染的优点。

迷宫机器人采用轮式结构，使用两轮差速插件模拟电动机驱动。

两轮差速插件是Gazebo中用于控制模型运动的插件。动态链接库gazebo_ros_diff_drive.so可以接收命令，设置模型的速度，还可以反馈模型的位置以及速度，如图2-2-2所示。

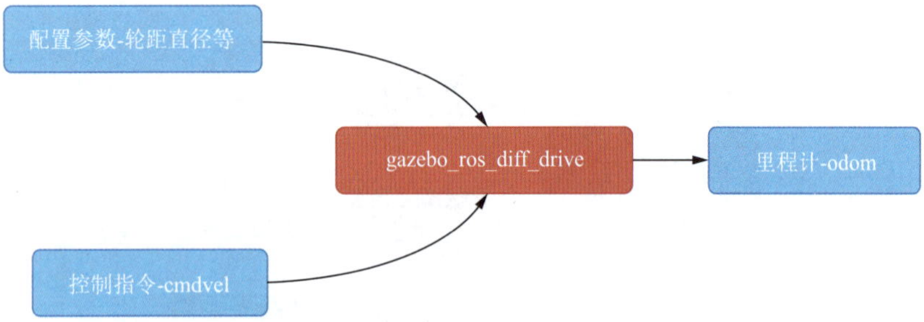

图2-2-2　两轮差速插件

两轮差速插件生效的前提是在urdf中正确导入插件，并配置相关参数。

两轮差速插件默认通过订阅话题/cmd_vel来获取目标线速度和目标角速度，该话题的消息类型为geometry_msgs/msg/Twist，可以向该话题发布数据来控制目标线速度和目标角速度。

两轮差速插件默认通过发布话题/odom来输出目标当前的位置、朝向（偏航角）和速度，该话题的消息类型为nav_msgs/msg/Odometry，可以订阅该话题来获取目标当前的位置、朝向（偏航角）和速度。

二、两轮差速插件添加

在urdf中添加两轮差速代码，并配置相关参数。

```xml
<gazebo>
    <plugin name='diff_drive' filename='libgazebo_ros_diff_drive.so'>
        <ros>
            <namespace>/</namespace> <!-- 命名空间-->
            <remapping>cmd_vel:=cmd_vel</remapping> <!-- 重映射话题名称-->
            <remapping>odom:=odom</remapping> <!-- 重映射话题名称-->
        </ros>
        <update_rate>60</update_rate><!-- 更新频率-->
        <!-- wheels -->
        <left_joint>left_wheel_joint</left_joint> <!-- left joint-->
        <right_joint>right_wheel_joint</right_joint> <!-- right joint-->
        <!-- kinematics -->
        <wheel_separation>0.07</wheel_separation> <!-- 两个轮子中心间距，单位m-->
        <wheel_diameter>0.02</wheel_diameter> <!-- 轮子的直径，单位m-->
        <!-- limits -->
        <max_wheel_torque>20</max_wheel_torque> <!-- 轮子的扭矩，单位Nm-->
        <max_wheel_acceleration>1.0</max_wheel_acceleration> <!-- 轮子的加速度-->
        <!-- output -->
        <publish_odom>true</publish_odom>  <!-- 是否发布odom话题-->
        <publish_odom_tf>true</publish_odom_tf>
        <publish_wheel_tf>false</publish_wheel_tf>
        <odometry_frame>odom</odometry_frame>
        <robot_base_frame>base_footprint</robot_base_frame>
    </plugin>
</gazebo>
```

三、两轮差速话题

添加两轮差速代码后，重新构建启动tqd.launch.py。

```
$ cd ~/Maze_robot_ws
$ colcon build
$ source install/setup.bash
$ ros2 launch tqd_Maze_robot tqd.launch.py
```

打印话题list。

```
$ ros2 topic list
```

终端中输出如下信息：

```
/camera_image
```

```
/camera_info
/clock
/cmd_vel                    # 设置目标线速度和目标角速度
/joint_states
/odom                       # 输出目标当前的位置、朝向（偏航角）和速度
/parameter_events
/rosout
/tf
```

1. /cmd_vel话题

（1）打印话题info

```
$ ros2 topic info /cmd_vel
```

终端中输出如下信息：

```
Type: geometry_msgs/msg/Twist
Publisher count: 0
Subscription count: 1
```

还没有发布者，所以**Publisher count**为0。

（2）打印话题echo

由于插件订阅**/cmd_vel**话题，所以在没有发布者的情况下，该话题没有echo消息。

（3）查看消息类型

```
$ ros2 interface show geometry_msgs/msg/Twist
```

终端中输出如下信息：

```
# 这表示在自由空间中的速度，分为线性部分和角度部分

Vector3  linear
    float64 x
    float64 y
    float64 z
Vector3  angular
    float64 x
    float64 y
    float64 z
```

2. /odom话题

（1）打印话题info

```
$ ros2 topic info /odom
```

终端中输出如下信息：

```
Type: sensor_msgs/msg/Odometry
Publisher count: 1
Subscription count: 0
```

还没有订阅者,所以Subscription count为0。

(2) 打印话题echo

```
$ ros2 topic echo /odom
```

终端中输出如下信息:

```
header:
  stamp:                                    # 本次发布的时间
    sec: 1459
    nanosec: 189000000
  frame_id: odom
child_frame_id: base_footprint
pose:                                       # pose信息
  pose:
    position:                               # 位置信息
      x: 0.0012131783619373808
      y: -2.584250942327375e-05
      z: 0.015999687192963755
    orientation:                            # 朝向信息
      x: 3.452774919449117e-07
      y: 7.119953618926764e-06
      z: -0.0005609542394105853
      w: 0.999999842639752
  covariance:                               # 矩阵
  - 1.0e-05
  - 0.0
  - 0.0
  - 0.0
  - 0.0
  - 0.0
  - 0.0
  - 1.0e-05
  - 0.0
  - 0.0
  - 0.0
  - 0.0
  - 0.0
  - 0.0
  - 1000000000000.0
  - 0.0
  - 0.0
  - 0.0
  - 0.0
  - 0.0
  - 0.0
  - 1000000000000.0
  - 0.0
  - 0.0
```

```yaml
      - 0.0
      - 0.0
      - 0.0
      - 0.0
      - 1000000000000.0
      - 0.0
      - 0.0
      - 0.0
      - 0.0
      - 0.0
      - 0.001
  twist:                                              # 当前速度
    twist:
      linear:                                         # 当前线速度
        x: -6.892120916898483e-07
        y: -1.2969665645975879e-07
        z: 0.0
      angular:                                        # 当前角速度
        x: 0.0
        y: 0.0
        z: -6.453089008299199e-06
    covariance:
      - 1.0e-05
      - 0.0
      - 0.0
      - 0.0
      - 0.0
      - 0.0
      - 1.0e-05
      - 0.0
      - 0.0
      - 0.0
      - 0.0
      - 0.0
      - 1000000000000.0
      - 0.0
      - 0.0
      - 0.0
      - 0.0
      - 0.0
      - 1000000000000.0
      - 0.0
      - 0.0
      - 0.0
      - 0.0
      - 0.0
      - 1000000000000.0
      - 0.0
      - 0.0
      - 0.0
```

```
        - 0.0
        - 0.0
        - 0.0
        - 0.001
```

（3）打印消息类型

```
$ ros2 interface show nav_msgs/msg/Odometry
```

终端中输出如下信息：

```
# 这表示对自由空间中的位置和速度的估计
# 此消息中的位置姿态应根据 header.frame_id 指定的坐标系来确定
# 此消息中的旋转应根据 child_frame_id 指定的坐标系来确定

# 包含父姿态的帧id
std_msgs/Header header
        builtin_interfaces/Time stamp
                int32 sec
                uint32 nanosec
        string frame_id

string child_frame_id

geometry_msgs/PoseWithCovariance pose
        Pose pose
                Point position
                        float64 x
                        float64 y
                        float64 z
                Quaternion orientation
                        float64 x 0
                        float64 y 0
                        float64 z 0
                        float64 w 1
        float64[36] covariance

geometry_msgs/TwistWithCovariance twist
        Twist twist
                Vector3  linear
                        float64 x
                        float64 y
                        float64 z
                Vector3  angular
                        float64 x
                        float64 y
                        float64 z
        float64[36] covariance
```

四、两轮差速程序控制

下面使用Python程序与差速插件进行交互。由于是在外部手动启动Python程序，所以不需要在tqd_Maze robot 中创建节点，直接在主目录中创建scripts目录放置练习脚本文件。

1. /cmd_vel话题

在scripts目录中创建diff_drive_cmd_vel.py，使用Python程序发布数据到/cmd_vel话题，为迷宫机器人设置目标速度。

```python
import rclpy
from rclpy.node import Node
from geometry_msgs.msg import Twist


def set_speed():
    rclpy.init(args=None)                # 初始化
    node = Node('publisher')             # 创建node
    # 创建发布者，话题/cmd_vel
    publisher = node.create_publisher(Twist, '/cmd_vel', 10)
    msg = Twist()                        # 消息类型

    def set_speed_callback():
        msg.linear.x = 0.1               # x线速度赋值
        msg.angular.z = 0.0              # z角速度赋值
        publisher.publish(msg)

    node.create_timer(0.5, set_speed_callback)
    rclpy.spin(node)        # 执行回调，执行这句程序后，ROS2才会开始工作
    rclpy.shutdown()        # 程序结束时（例如Ctrl+C），关闭ROS2节点

set_speed()
```

运行程序后，迷宫机器人将沿着直线行走，如图2-2-3所示。尝试修改msg.linear.x和msg.angular.z，观察运行状态和运行轨迹。

2. /odom话题

在scripts目录中创建diff_drive_odom.py，使用Python程序订阅/odom话题，获取迷宫机器人在Gazebo中的位置和偏航角。

在ROS1中，集成了用于坐标转换的tf模块，而在ROS2中，需要自行安装：

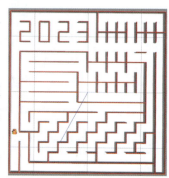

图2-2-3 驱动迷宫机器人运动

```
$ sudo apt install python3-pip       # 安装pip，方便管理python包
$ pip install transforms3d
$ sudo apt install ros-humble-tf-transformations

import rclpy
from rclpy.node import Node
from nav_msgs.msg import Odometry
import tf_transformations


def get_odom_callback(msg):
    position = msg.pose.pose.position
    quaternion = (
        msg.pose.pose.orientation.x,
        msg.pose.pose.orientation.y,
        msg.pose.pose.orientation.z,
        msg.pose.pose.orientation.w)
    euler = tf_transformations.euler_from_quaternion(quaternion)
                                          # 将四元数转为欧拉角
    print(position.x, position.y)         # 机器人在Gazebo中的坐标
    print(euler[2])                       # 偏航角，-3.14 ～ 3.14

def get_odom():
    rclpy.init(args=None)
node = Node('subscriber')
# 创建订阅者，话题/odom，回调get_odom_callback
    node.create_subscription(Odometry, '/odom', get_odom_callback, 10)
    rclpy.spin(node)
    rclpy.shutdown()

get_odom()
```

运行程序，移动或旋转迷宫机器人，观察输出的数据，如图2-2-4、图2-2-5所示。

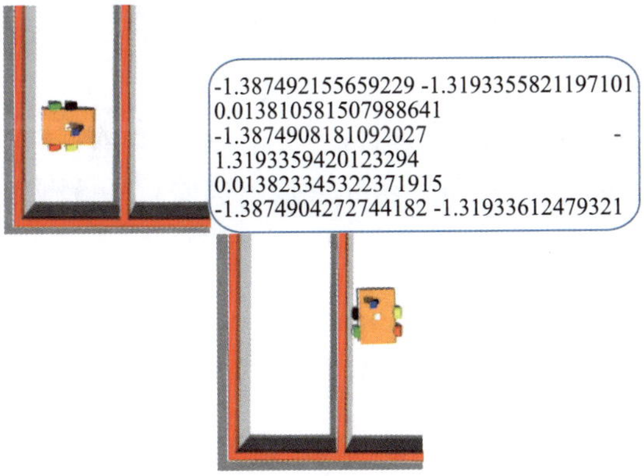

-1.387492155659229 -1.3193355821197101
0.013810581507988641
-1.3874908181092027 -1.3193359420123294
0.013823345322371915
-1.3874904272744182 -1.31933612479321

-1.207654964537468 -1.2368954466120166
1.6050459071286887
-1.2076550248187337 -1.2368947923508726
1.6050461289656925
-1.2076551376432874 -1.2368941396893731
1.60504644409013734

图2-2-4　位置一 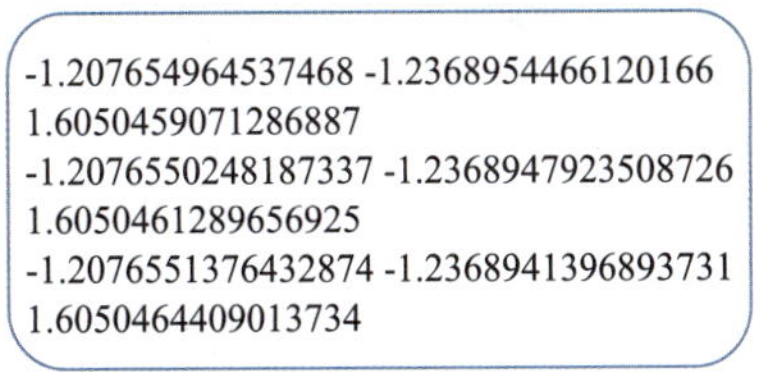　　　　　　　　　　图2-2-5　位置二

任务二　学习激光雷达检测

本任务目的是学习使用激光雷达插件，检测迷宫机器人周围的障碍物距离。

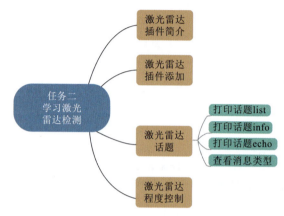

一、激光雷达插件简介

普通的单线激光雷达一般有一个发射器，一个接收器，发射器发出激光射线到前方的目标上，物体会将激光反射回来，然后激光雷达的接收器可以检测到反射的激光，如图2-2-6所示。

通过计算发送和反馈之间的时间间隔，再乘以激光的速度，就可以计算出激光飞行的距离（d）。

激光雷达插件是Gazebo中用于检测障碍物距离的插件。动态链接库libgazebo_ros_ray_sensor.so包含了激光雷达和超声波两种光线插件。可以接收命令，设置模型的速度，还可以反馈模型的位置以及速度。

图2-2-6　激光检测示意图

激光雷达插件生效的前提是在urdf中正确导入插件，并配置相关参数。

激光雷达插件默认通过发布话题/<pluginname>/out来输出当前激光检测到的距离信息，可以选择重映射，例如/scan。

该话题的消息类型为sensor_msgs/msg/LaserScan。

可以订阅该话题来获取当前激光检测到的距离信息。

二、激光雷达插件添加

在urdf中添加激光雷达代码，并配置相关参数。

```xml
<gazebo reference="laser_link">         <!-- 指定生效的link-->
    <sensor name="laser_sensor" type="ray">   <!-- 指定name和type-->
        <pose>0 0 0.075 0 0 0</pose>  <!-- pose位置,同joint-->
        <always_on>true</always_on>         <!-- 常开-->
        <visualize>true</visualize>    <!-- 蓝色光线是否可见-->
        <update_rate>60</update_rate>   <!-- 更新频率-->
        <ray>
            <scan>              <!-- 定义扫描角度范围-->
                <horizontal>    <!-- horizontal: 水平扫描, vertical: 垂直扫描-->
                    <samples>360</samples>    <!-- 取样数量, int-->
                    <resolution>1.000000</resolution>  <!-- 分辨率-->
                    <min_angle>0.000000</min_angle> <!-- 扫描起始角度-->
                    <max_angle>6.280000</max_angle> <!-- 扫描终止角度-->
                </horizontal>
            </scan>
            <range>              <!-- 定义扫描距离范围-->
                <min>0.120000</min>   <!-- 最小距离-->
                <max>3.5</max>         <!-- 最大距离-->
                <resolution>0.015000</resolution>   <!-- 分辨率-->
            </range>
        </ray>
        <plugin name="laserscan" filename="libgazebo_ros_ray_sensor.so">  <!-- 指定插件名-->
            <ros>
                <remapping>~/out:=scan</remapping>  <!-- 重映射话题名称-->
            </ros>
            <output_type>sensor_msgs/LaserScan</output_type> <!-- 输出的消息类型-->
            <frame_name>laser_link</frame_name>
        </plugin>
    </sensor>
</gazebo>
```

三、激光雷达话题

添加激光雷达代码后,重新构建启动tqd.launch.py。

```
$ cd ~/Maze_robot_ws
$ colcon build
$ source install/setup.bash
$ ros2 launch tqd_Maze_robot  tqd.launch.py
```

启动后，可以看到蓝色的区域就是激光雷达扫描的区域，如图2-2-7所示。

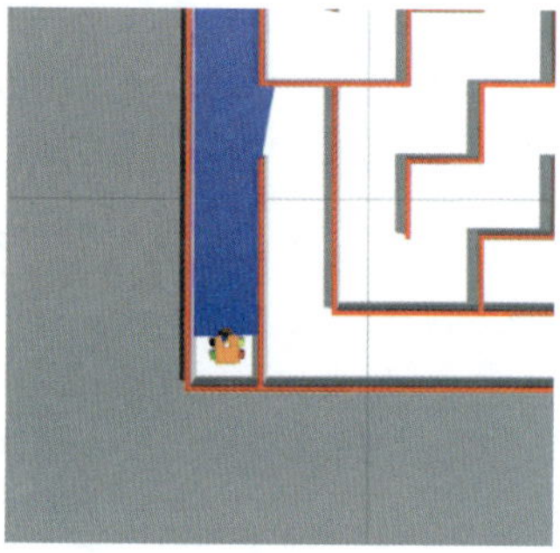

1. 打印话题list

```
$ ros2 topic list
```

终端中输出如下信息：

```
/camera_image
/camera_info
/clock
/cmd_vel
/joint_states
/odom
/parameter_events
/rosout
/scan                          # 激光雷达话题
/tf
```

图2-2-7 激光雷达扫描的区域

2. 打印话题info

```
$ ros2 topic info /scan
```

终端中输出如下信息：

```
Type: sensor_msgs/msg/LaserScan
Publisher count: 1
Subscription count: 0
```

还没有订阅者，所以Subscription count为0。

3. 打印话题echo

```
$ ros2 topic echo /scan
```

终端中输出如下信息：

```
header:
  stamp:
    sec: 1280
    nanosec: 57000000
  frame_id: laser_link
angle_min: -1.5708999633789062
angle_max: 1.5708999633789062
angle_increment: 0.008751532062888145
time_increment: 0.0
scan_time: 0.0
range_min: 0.019999999552965164
range_max: 3.5
```

```
ranges:
- 0.09237094968557358
- 0.09239806979894638
- 0.09243228286504745
- 0.09247361123561859
- 0.09252206236124039
...
```

4. 查看消息类型

```
$ ros2 interface show sensor_msgs/msg/LaserScan
```

终端中输出如下信息：

```
# Single scan from a planar laser range-finder
#
std_msgs/Header header
        builtin_interfaces/Time stamp
                int32 sec
                uint32 nanosec
        string frame_id
float32 angle_min
float32 angle_max
float32 angle_increment
float32 time_increment
float32 scan_time
float32 range_min
float32 range_max
float32[] ranges
float32[] intensities
```

四、激光雷达程序控制

下面使用Python程序与激光雷达插件进行交互。由于是在外部手动启动Python程序，所以不需要在tqd_Maze robot中创建节点，直接在scripts目录中创建脚本文件ray_sensor.py。

使用Python程序订阅/scan话题，获取Gazebo中迷宫机器人周围的障碍物距离。

视　频

激光雷达检测

```
import rclpy
from rclpy.node import Node
from sensor_msgs.msg import LaserScan

def get_laser_callback(msg):
    region = msg.ranges
    # 输出 水平右侧、右斜45、正前方、左斜45、水平左侧 距离
    print('right:', region[5])
    print('rf45 :', region[90])
```

```
        print('front:', region[180])
        print('lf45 :', region[270])
    print('left :', region[355])

def get_laser():
    rclpy.init(args=None)
node_laser = Node('laser')
# 创建订阅者，话题/scan，回调get_laser_callback
    node_laser.create_subscription(LaserScan, '/scan', get_laser_callback, 10)
    rclpy.spin(node_laser)
    rclpy.shutdown()

get_laser()
```

运行程序，移动或旋转迷宫机器人，观察输出的数据，如图2-2-8、图2-2-9、图2-2-10所示。

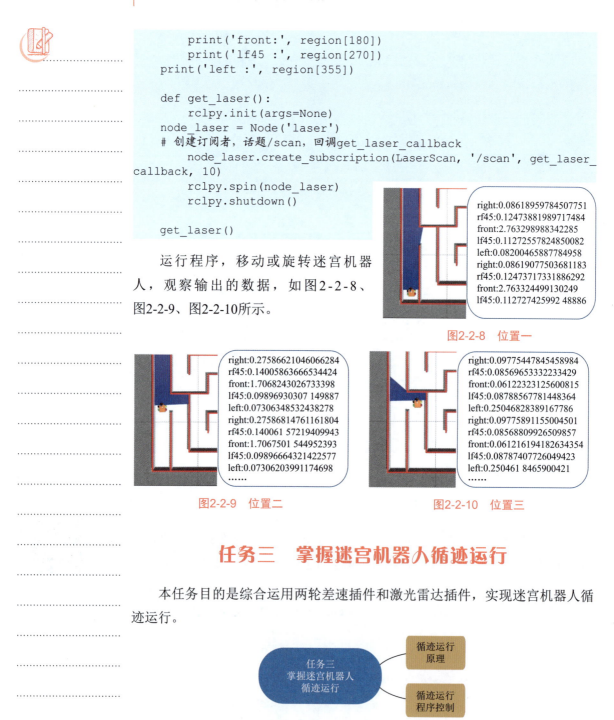

图2-2-8　位置一

图2-2-9　位置二　　　　　　　　图2-2-10　位置三

任务三　掌握迷宫机器人循迹运行

本任务目的是综合运用两轮差速插件和激光雷达插件，实现迷宫机器人循迹运行。

一、循迹运行原理

迷宫机器人最佳的运行轨迹是沿着迷宫中心线行走，这样距离两侧挡板有相同的间距,保证迷宫机器人能够顺利到达迷宫终点,如图2-2-11所示。

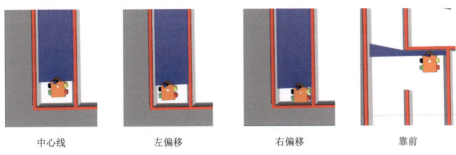

图2-2-11 位置示意图

激光雷达插件可以检测距左侧、前方和右侧挡板的距离,判断迷宫机器人在一个单元格中的位置。

两轮差速插件可以控制迷宫机器人运动的线速度和角速度,直线直行和转弯,如图2-2-12所示。

依据激光雷达插件检测到的挡板距离,调整两轮差速插件设置的目标速度,就可以实现迷宫机器人的循迹运行。

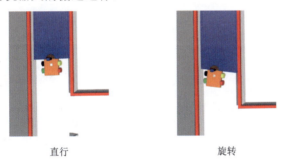

图2-2-12 动作示意图

二、循迹运行程序控制

在scripts目录中创建脚本文件track.py。

使用Python程序订阅/scan话题,获取Gazebo中迷宫机器人周围的障碍物距离。

使用Python程序发布数据到/cmd_vel话题,为迷宫机器人设置目标速度,如图2-2-13所示。

由本项目任务二可知,当迷宫机器人位于迷宫中心线上时,水平左侧和右侧的挡板距离是0.085 m左右,所以在程序中使用0.085作为阈值,判断是否发生了偏移。

视 频

循迹运行

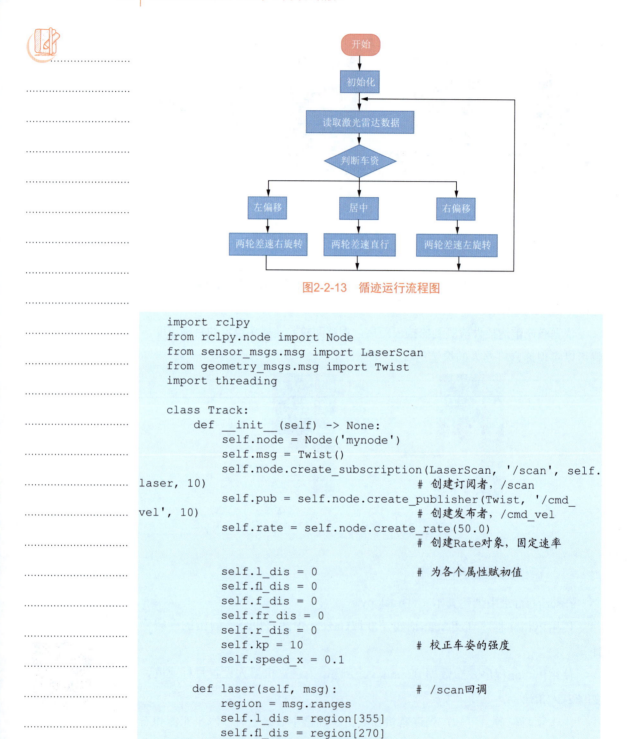

图2-2-13 循迹运行流程图

```
import rclpy
from rclpy.node import Node
from sensor_msgs.msg import LaserScan
from geometry_msgs.msg import Twist
import threading

class Track:
    def __init__(self) -> None:
        self.node = Node('mynode')
        self.msg = Twist()
        self.node.create_subscription(LaserScan, '/scan', self.laser, 10)                      # 创建订阅者,/scan
        self.pub = self.node.create_publisher(Twist, '/cmd_vel', 10)                           # 创建发布者,/cmd_vel
        self.rate = self.node.create_rate(50.0)
                                                # 创建Rate对象,固定速率

        self.l_dis = 0                          # 为各个属性赋初值
        self.fl_dis = 0
        self.f_dis = 0
        self.fr_dis = 0
        self.r_dis = 0
        self.kp = 10                            # 校正车姿的强度
        self.speed_x = 0.1

    def laser(self, msg):                       # /scan回调
        region = msg.ranges
        self.l_dis = region[355]
        self.fl_dis = region[270]
        self.f_dis = region[180]
        self.fr_dis = region[90]
        self.r_dis = region[5]
```

```python
    def ros_spin(self):                    # rclpy.spin(), 在子线程调用
        rclpy.spin(self.node)

    def posture_adjust(self):              # 判断是否偏移, 然后校正
        if self.l_dis<0.085:
            speed_z_temp = 0.5*(self.l_dis-0.085)*self.kp
        elif self.r_dis<0.085:
            speed_z_temp = -0.5*(self.r_dis-0.085)*self.kp
        else:
            speed_z_temp = 0.0
        return speed_z_temp

    def move(self):                        # 迷宫机器人运行
        while rclpy.ok():
            self.msg.linear.x = self.speed_x
            self.msg.angular.z = self.posture_adjust()
            self.pub.publish(self.msg)
            self.rate.sleep()              # 实现固定速率发布
            if 0.05<self.f_dis<0.08:# 检测到前方挡板时停车
                self.msg.linear.x = 0.0
                self.msg.angular.z = 0.0
                self.pub.publish(self.msg)
                self.rate.sleep()
                break

def main():
    rclpy.init(args=None)
    track = Track()
    t = threading.Thread(None, target=track.ros_spin, daemon=True)
                                           # 在子线程调用rclpy.spin()
    t.start()
    track.move()

if __name__ == '__main__':
    main()
```

思考与总结

1. 如何协调控制模型的驱动速度和实际速度?

2. 如何调节激光雷达检测的角度范围和距离范围?

3. 在使用水平传感器循迹运行时,经常会发现无法及时校正的情况,应当如何改进?

4. 在程序中可以读取和控制多个话题,从而实现对模型的姿态检测和运动控制。

项目三

掌握迷宫机器人智能控制

学习目标

1. 掌握迷宫机器人在迷宫中坐标的计算方法。
2. 掌握激光雷达检测数据的处理方法。
3. 掌握迷宫机器人在迷宫中精确转弯的控制方法。

任务一　学习坐标计算

本任务目的是依据迷宫机器人在Gazebo中的位置，计算在迷宫中的坐标。

一、坐标计算原理

迷宫机器人的基本功能是从起点运行到终点，那么迷宫机器人如何确定自身在迷宫中的哪个位置呢？答案是根据迷宫坐标。

在迷宫中，每个单元格均用一个坐标表示，起点坐标定义为（0，0），右上角定义为（F，F），如图2-3-1所示。

迷宫机器人在迷宫中的坐标有两种计算方式：

方式一：根据自身在Gazebo中的位置，确定属于哪一个单元格，

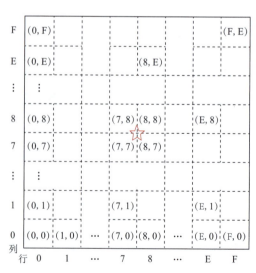

图2-3-1　迷宫坐标示意图

从而确定迷宫坐标。

方式二：根据自身在Gazebo中移动的方向和距离，计算迷宫坐标。

方式一是通过使用绝对意义上的位置来计算坐标，但由于Gazebo是一个仿真的环境，随着时间的推移，模型会发生移动，所以迷宫机器人在Gazebo中的绝对位置并不准确。因此通常采用方式二来计算迷宫坐标。

两轮差速插件可以获取迷宫机器人在Gazebo中的位置，并驱动迷宫机器人运行，可以用来计算迷宫坐标。步骤如下：

① 在驱动迷宫机器人前记录其在Gazebo中的位置。
② 驱动迷宫机器人开始运行。
③ 每当迷宫机器人移动距离超过一个单元格时，在移动的方向上坐标+1。

确定迷宫机器人的起始坐标后，循环以上三个步骤，就可以计算出新的坐标，如图2-3-2所示。

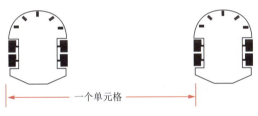

图2-3-2　坐标更新示意图

二、坐标计算程序控制

1. 启动tqd_Maze robot

首先启动tqd_Maze robot 功能包中的tqd.launch.py。

```
$ cd ~/Maze robot _ws
$ source install/setup.bash
$ ros2 launch tqd_Maze robot  tqd.launch.py
```

2. 创建coordinate.py

在scripts目录中创建脚本文件coordinate.py。编写如下代码：

```
import rclpy
from rclpy.node import Node
from std_msgs.msg import String
from geometry_msgs.msg import Twist
from nav_msgs.msg import Odometry
from sensor_msgs.msg import LaserScan
import tf_transformations
```

视频

坐标计算

```python
import threading

class Drive(object):
    def __init__(self) -> None:
        self.valueinit()
        self.ros_register()

    def valueinit(self):
        self.l_dis = 0
        self.fl_dis = 0
        self.f_dis = 0
        self.fr_dis = 0
        self.r_dis = 0
        self.position_x = 0
        self.position_y = 0
        self.yaw = 1.5708

        self.blocksize = 0.178        # 迷宫单元格预设大小
        self.kp = 10

        self.speed_x = 0.10
        self.speed_z = 0.35

    def ros_register(self):
        self.msg = Twist()
        self.string = String()
        self.node = Node('mynode')
        self.node.create_subscription(LaserScan, '/scan', self.laser, 10)                                    # 创建订阅者，/scan
        self.node.create_subscription(Odometry, '/odom', self.odom, 10)                                     # 创建订阅者，/odom
        self.pub = self.node.create_publisher(Twist, '/cmd_vel', 10)
                                              # 创建发布者，/cmd_vel
        self.rate = self.node.create_rate(50.0)

    def ros_spin(self):              # rclpy.spin()，在子线程调用
        rclpy.spin(self.node)

    def laser(self, msg):            # /scan回调
        region = msg.ranges
        self.l_dis = region[340]
        self.fl_dis = region[270]
        self.f_dis = region[180]
        self.fr_dis = region[90]
        self.r_dis = region[20]

    def get_laser(self):
```

```python
        return self.l_dis, self.fl_dis, self.f_dis, self.fr_dis, self.r_dis

    def odom(self, msg):                    # /odom回调
        position = msg.pose.pose.position
        self.position_x = position.x
        self.position_y = position.y
        quaternion = (
            msg.pose.pose.orientation.x,
            msg.pose.pose.orientation.y,
            msg.pose.pose.orientation.z,
            msg.pose.pose.orientation.w)
        euler = tf_transformations.euler_from_quaternion(quaternion)
        self.yaw = euler[2]

    def get_odom(self):
        return self.position_x, self.position_y, self.yaw

    def posture_adjust(self):               # 判断是否偏移，然后校正
        if self.f_dis>0.09:
            if self.fl_dis<0.118:
                speed_z_temp = 0.5*(self.fl_dis-0.118)*self.kp
            elif self.fr_dis<0.118:
                speed_z_temp = -0.5*(self.fr_dis-0.118)*self.kp
            else:
                speed_z_temp = 0.0
        else:
            if self.l_dis<0.085:
                speed_z_temp = 0.5*(self.l_dis-0.085)*self.kp
            elif self.r_dis<0.085:
                speed_z_temp = -0.5*(self.r_dis-0.085)*self.kp
            else:
                speed_z_temp = 0.0
        return speed_z_temp

    def move(self, numblock=1):             # 驱动迷宫机器人运行
        flag = 1
        while rclpy.ok():
            self.msg.linear.x = self.speed_x
            self.msg.angular.z = self.posture_adjust()
            self.pub.publish(self.msg)
            self.rate.sleep()
            if flag:      # 刚开始驱动时self.get_odom()返回的是0，因此
                          # 添加此判断
                tempx, tempy, tempz = self.get_odom()
```

```
                        if (not tempx) and (not tempy):
                            continue
                        flag = 0
                    if abs(self.get_odom()[0] - tempx)>=self.blocksize*
numblock or abs(self.get_odom()[1] - tempy)>=self.blocksize*numblock:
                        # 当移动的距离超过设定值时结束循环
                        self.msg.linear.x = 0.0
                        self.msg.angular.z = 0.0
                        self.pub.publish(self.msg)
                        self.rate.sleep()
                        break


def main():
    rclpy.init(args=None)
    drive = Drive()
    t = threading.Thread(None, target=drive.ros_spin, daemon=True)
    t.start()
    drive.move(10)        # 移动10个单元格

if __name__ == '__main__':
    try:
        main()
    finally:
        rclpy.shutdown()
```

修改drive.move(10)行走的单元格数量，观察在迷宫中实际运行的单元格数量，若不匹配，如图2-3-3所示，适当修改self.blocksize的大小，直至两者匹配，如图2-3-4所示。

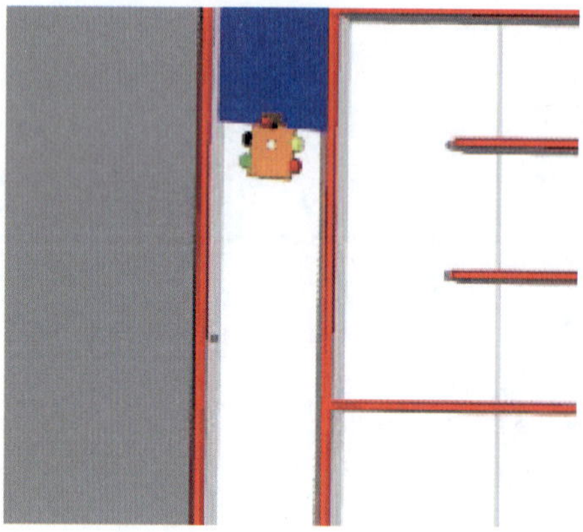

图2-3-3 不匹配 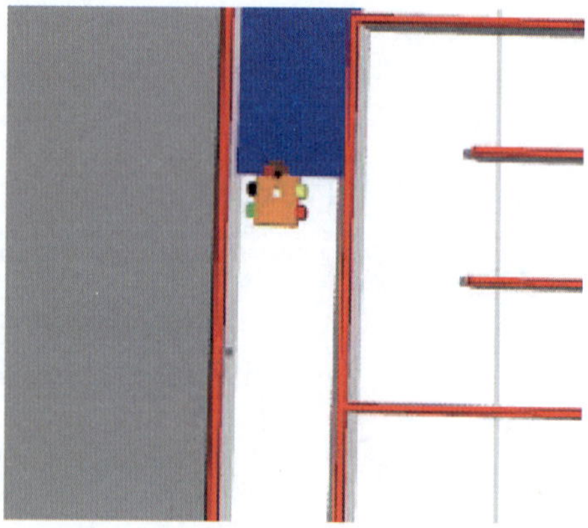　　　　　　　图2-3-4 匹配

任务二 学习路口检测

本任务目的是依据激光雷达检测的数据，判断当前位置的墙壁资料，即左侧、前方和右侧分别是墙壁还是路口。

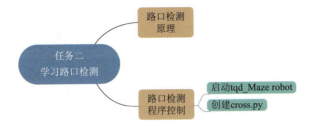

一、路口检测原理

迷宫机器人使用激光雷达检测四周挡板的距离，其检测距离可以在urdf模型中指定，为了提高检测精度和方便扩展，检测下限通常会选择0.02 m左右，过大或过小都可能会造成误检测；检测上限通常会选择大于1 m的数值，太小将不利于后期程序优化。

迷宫机器人在单元格中行走时，最完美的运行轨迹是沿着中心线行走。如果发生了偏移，最极端的情况是紧贴左侧挡板或者紧贴右侧挡板，如图2-3-5所示。

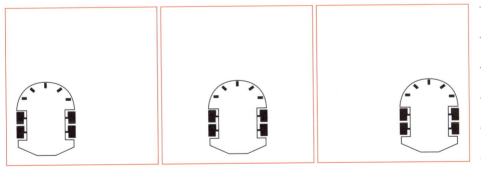

图2-3-5 迷宫机器人位置示意图

迷宫机器人在单元格正中间时，距离两侧挡板的阈值近似是0.08 m；当迷宫机器人紧贴左侧或紧贴右侧时，距离另一侧挡板的距离约为0.10 m，如图2-3-6所示。

因此，当迷宫机器人检测的左侧、前方和右侧墙壁距离大于0.12时就可以断定单元格在该方向上存在路口，如图2-3-7所示。

结合迷宫机器人的坐标，就可以将所有经过的坐标的墙壁资料都保存下来。

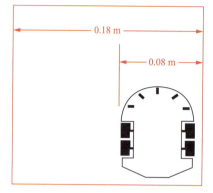

图2-3-6　单元格与机器人尺寸示意图

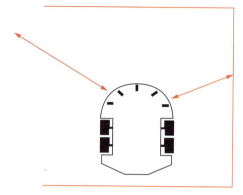

图2-3-7　路口示意图

二、路口检测程序控制

1. 启动tqd_Maze robot

首先启动tqd_Maze robot 功能包中的tqd.launch.py。

```
$ cd ~/Maze robot _ws
$ source install/setup.bash
$ ros2 launch tqd_Maze robot  tqd.launch.py
```

2. 创建cross.py

在scripts目录中创建脚本文件cross.py。复制本项目任务一中的代码，修改main程序，如下所示：

```python
def main():
    rclpy.init(args=None)
    drive = Drive()
    t = threading.Thread(None, target=drive.ros_spin, daemon=True)
    t.start()
    x, y = 0, 0
    while 1:
        drive.move()
        y = y+1
        l_dis, fl_dis, f_dis, fr_dis, r_dis = drive.get_laser()
        leftpath = '1' if l_dis>0.12 else '0'
        backpath = '1'
        rightpath = '1' if r_dis>0.12 else '0'
        frontpath = '1' if f_dis>0.12 else '0'
        print(f"({x}, {y}), {leftpath} {backpath} {rightpath} {frontpath}")
        if y==10:
            break
```

运行程序，观察输出的数据，如图2-3-8所示。

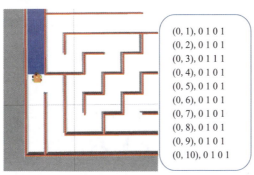

图2-3-8　实时判断单元格墙壁信息

还可以尝试将以下语句注释掉，运行程序，移动迷宫机器人，观察输出的数据，如图2-3-9、图2-3-10所示。

```
# drive.move()
# y = y+1
```

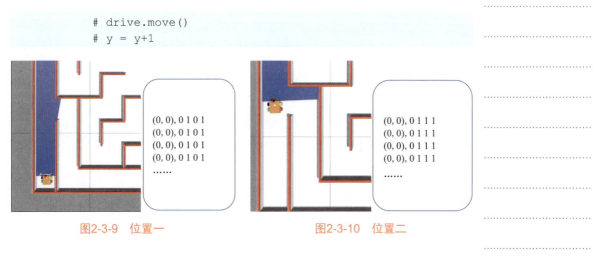

图2-3-9　位置一　　　　　　　　　图2-3-10　位置二

任务三　掌握转弯控制

本任务目的是依据偏航角的改变，控制迷宫机器人进行精确的转弯。

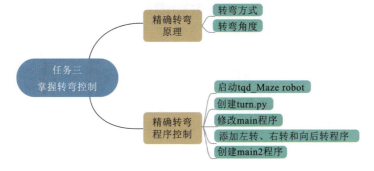

一、精确转弯原理

迷宫机器人在迷宫中运行时除直行外，还会遇到大量转弯。转弯是否准确，对迷宫机器人的后续运行影响巨大。精确转弯需要考虑两方面的因素，转弯方式和转弯角度。

1. 转弯方式

迷宫机器人的常用转弯方式有三种，如图2-3-11所示。

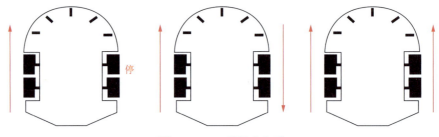

图2-3-11 三种转弯方式

方式一：内侧轮停止，外侧轮正转，这种转弯方式等效于以内侧轮为圆心，转动1/4圆弧，转弯半径较大。

方式二：内侧轮反转，外侧轮正转，且两轮速度相同，这种转弯方式等效于以迷宫机器人中心为圆心，转动1/4圆弧，转弯半径最小，稳定性最高。

方式三：内侧轮和外侧轮均正转，但外侧轮速度较高，这种转弯方式速度最快，但控制难度稍大。

接下来采用方式二作为迷宫机器人的转弯方式。

2. 转弯角度

转弯角度的大小主要取决于何时停止转弯，如图2-3-12所示，迷宫机器人的转弯角度控制主要采用两种方式：

方式一：延时控制。在驱动迷宫机器人进入转动状态后，延时一定的时间，然后再结束转动。这种控制方式最简单，但精度较差。

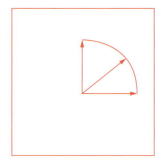

图2-3-12 转弯角度示意图

方式二：传感器检测控制。通过使用传感器检测迷宫机器人的姿态变化，当达到预设值时，结束转动。这种控制方式依赖传感器检测，精度较高。

在Gazebo仿真中，有多种传感器可用于检测迷宫机器人的姿态变化，如两轮差速插件提供的偏航角，如图2-3-13所示，图中yaw表示偏航角度。

迷宫机器人车头向左时，存在π到-π的跳变，为了方便计算，对偏航角进行修正，将所有的偏航角数据增加π，如图2-3-14所示。在程序中，π按照3.141 59进行计算。

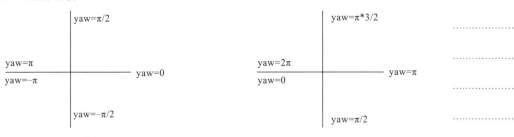

图2-3-13　偏航角示意图　　　图2-3-14　偏航角修正

下面就选择偏航角作为转弯角度的控制方式。

二、精确转弯程序控制

1. 启动tqd_Maze robot

首先启动tqd_Maze robot 功能包中的tqd.launch.py。

```
$ cd ~/Maze robot_ws
$ source install/setup.bash
$ ros2 launch tqd_Maze robot  tqd.launch.py
```

2. 创建turn.py

在scripts目录中创建脚本文件turn.py。复制本项目任务一中的代码，在valueinit()方法中增加对偏航角的赋值。

```
def valueinit(self):
    self.l_dis = 0
    self.fl_dis = 0
    self.f_dis = 0
    self.fr_dis = 0
    self.r_dis = 0
    self.position_x = 0
    self.position_y = 0
    self.yaw = 0
    self.fyaw = 4.71            # 车头向前时的偏航角
    self.ryaw = 3.14            # 车头向右时的偏航角
    self.byaw = 1.57            # 车头向下时的偏航角
    self.lyaw0 = 0              # 车头向左时的偏航角0
    self.lyaw1 = 6.28           # 车头向左时的偏航角1

    self.blocksize = 0.175
    self.kp = 10
```

```
        self.speed_x = 0.10
        self.speed_z = 0.35
```

3. 修改main程序

```
def main():
    rclpy.init(args=None)
    drive = Drive()
    t = threading.Thread(None, target=drive.ros_spin, daemon=True)
    t.start()
    x, y = 0, 0
    while 1:
        drive.move()
        y = y+1
        l_dis, fl_dis, f_dis, fr_dis, r_dis = drive.get_laser()
        leftpath = '1' if l_dis>0.12 else '0'
        backpath = '1'
        rightpath = '1' if r_dis>0.12 else '0'
        frontpath = '1' if f_dis>0.12 else '0'

        if rightpath=='1':
            drive.turn(0)
            break

        if y==10:
            break
```

运行程序，当迷宫机器人检测到右侧有路口时，右转弯，然后停车。

4. 添加左转、右转和向后转程序

在迷宫机器人搜索或冲刺时，涉及的动作是左转、右转和向后转，因此还需要深入判断当前yaw是哪个方向，需要转动到哪个方向，添加右转、左转和向后转代码。

```
    def turnright(self):
        flag = 1
        while rclpy.ok():
            self.msg.linear.x = 0.0
            self.msg.angular.z = -self.speed_z
            self.pub.publish(self.msg)
            self.rate.sleep()
            if flag:
                oldyaw = self.get_odom()[2]
                if not oldyaw:
                    continue
                flag = 0
```

```python
            if 4.41<oldyaw<5.01:
                if self.ryaw-0.1<self.yaw<self.ryaw+0.1:
                    self.msg.linear.x = 0.0
                    self.msg.angular.z = 0.0
                    self.pub.publish(self.msg)
                    # rclpy.spin_once(self.node)
                    self.rate.sleep()
                    break

            elif 2.84<oldyaw<3.44:
                if self.byaw-0.1<self.yaw<self.byaw+0.1:
                    self.msg.linear.x = 0.0
                    self.msg.angular.z = 0.0
                    self.pub.publish(self.msg)
                    # rclpy.spin_once(self.node)
                    self.rate.sleep()
                    break

            elif 1.27<oldyaw<1.87:
                if self.yaw<self.lyaw0+0.1:
                    self.msg.linear.x = 0.0
                    self.msg.angular.z = 0.0
                    self.pub.publish(self.msg)
                    # rclpy.spin_once(self.node)
                    self.rate.sleep()
                    break

            elif oldyaw<0.3 or oldyaw>5.98:
                if self.fyaw-0.1<self.yaw<self.fyaw+0.1:
                    self.msg.linear.x = 0.0
                    self.msg.angular.z = 0.0
                    self.pub.publish(self.msg)
                    # rclpy.spin_once(self.node)
                    self.rate.sleep()
                    break

    def turnleft(self):
        flag = 1
        while rclpy.ok():
            self.msg.linear.x = 0.0
            self.msg.angular.z = self.speed_z
            self.pub.publish(self.msg)
            self.rate.sleep()
            if flag:
                oldyaw = self.yaw
                if not oldyaw:
```

```python
                    continue
                flag = 0
            if 4.41<oldyaw<5.01:
                if self.yaw>self.lyaw1-0.1:
                    self.msg.linear.x = 0.0
                    self.msg.angular.z = 0.0
                    self.pub.publish(self.msg)
                    # rclpy.spin_once(self.node)
                    self.rate.sleep()
                    break
            elif 2.84<oldyaw<3.44:
                if self.fyaw-0.1<self.yaw<self.fyaw+0.1:
                    self.msg.linear.x = 0.0
                    self.msg.angular.z = 0.0
                    self.pub.publish(self.msg)
                    # rclpy.spin_once(self.node)
                    self.rate.sleep()
                    break
            elif 1.27<oldyaw<1.87:
                if self.ryaw-0.1<self.yaw<self.ryaw+0.1:
                    self.msg.linear.x = 0.0
                    self.msg.angular.z = 0.0
                    self.pub.publish(self.msg)
                    # rclpy.spin_once(self.node)
                    self.rate.sleep()
                    break
            elif oldyaw<0.3 or oldyaw>5.98:
                if self.byaw-0.1<self.yaw<self.byaw+0.1:
                    self.msg.linear.x = 0.0
                    self.msg.angular.z = 0.0
                    self.pub.publish(self.msg)
                    # rclpy.spin_once(self.node)
                    self.rate.sleep()
                    break

    def turnback(self):
        flag = 1
        while rclpy.ok():
            self.msg.linear.x = 0.0
            self.msg.angular.z = -self.speed_z
            self.pub.publish(self.msg)
            self.rate.sleep()
            if flag:
                oldyaw = self.get_odom()[2]
                if not oldyaw:
                    continue
                flag = 0
```

```
                if 4.41<oldyaw<5.01:
                    if self.byaw-0.1<self.yaw<self.byaw+0.1:
                        self.msg.linear.x = 0.0
                        self.msg.angular.z = 0.0
                        self.pub.publish(self.msg)
                        # rclpy.spin_once(self.node)
                        self.rate.sleep()
                        break
                elif 2.84<oldyaw<3.44:
                    if self.yaw<self.lyaw0+0.1:
                        self.msg.linear.x = 0.0
                        self.msg.angular.z = 0.0
                        self.pub.publish(self.msg)
                        # rclpy.spin_once(self.node)
                        self.rate.sleep()
                        break
                elif 1.27<oldyaw<1.87:
                    if self.fyaw-0.1<self.yaw<self.fyaw+0.1:
                        self.msg.linear.x = 0.0
                        self.msg.angular.z = 0.0
                        self.pub.publish(self.msg)
                        # rclpy.spin_once(self.node)
                        self.rate.sleep()
                        break
                elif oldyaw<0.3 or oldyaw>5.98:
                    if self.ryaw-0.1<self.yaw<self.ryaw+0.1:
                        self.msg.linear.x = 0.0
                        self.msg.angular.z = 0.0
                        self.pub.publish(self.msg)
                        # rclpy.spin_once(self.node)
                        self.rate.sleep()
                        break
```

5. 创建main2程序

```
def main2():
    rclpy.init(args=None)
    drive = Drive()
    t = threading.Thread(None, target=drive.ros_spin, daemon=True)
    t.start()
    x, y = 0, 0
    while 1:
        drive.move()
        y = y+1
        l_dis, fl_dis, f_dis, fr_dis, r_dis = drive.get_laser()
        leftpath = '1' if l_dis>0.12 else '0'
        backpath = '1'
```

```
            rightpath = '1' if r_dis>0.12 else '0'
            frontpath = '1' if f_dis>0.12 else '0'

            if rightpath=='1':
                drive.turnright()
                break

            if leftpath=='1':
                drive.turnleft()
                break

            if frontpath=='0':
                drive.turnback()
                break

            if y==10:
                break
```

执行main2程序，观察迷宫机器人的运行情况。还可以尝试移动迷宫机器人到其他位置，例如左侧有路口或前方有挡板的区域，再次执行程序，观察现象。

思考与总结

1. 如何提高墙壁的检测精度？

2. 坐标计算有哪些方式？各有什么优缺点？

3. 转弯的方式有多种，行进中转弯较快，但运动过程比较复杂；原地转弯较慢，但控制较简单。

第三篇 迷宫机器人仿真实践

在掌握迷宫机器人仿真的基本控制方法后，就可以尝试进行迷宫机器人仿真竞赛了。本篇主要讲解，在仿真环境中实现迷宫机器人的各种程序逻辑，如搜索策略、信息存取智能算法等；最后使用编写的程序进行一次仿真竞赛。

学习迷宫机器人智能算法

学习目标

1. 了解迷宫的搜索策略及其优缺点。
2. 了解迷宫信息的存储与读取方式。
3. 了解迷宫最优路径的规划方式。

任务一　了解迷宫搜索策略

本任务目的是了解迷宫机器人搜索迷宫的方式,掌握常用的搜索法则。

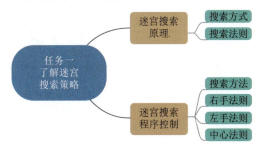

一、迷宫搜索原理

在第一次运行时,对于迷宫机器人来说,迷宫是未知的,不确定如何行走才能到达终点。因此需要去搜索迷宫,直到抵达终点为止。

1. 搜索方式

迷宫机器人搜索迷宫通常有两种方式:

方式一:尽快到达终点,如图3-1-1所示。

方式二:搜索整个迷宫,如图3-1-2所示。

这两种方式各有利弊。利用方式一虽然可以缩短搜索迷宫所需的时间,但不一定能够得到整个迷宫地图的资料。若找到的路径不是迷宫的最优路径,这将会影响迷宫机器人最后冲刺的时间。利用方式二,

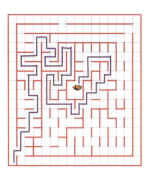

图3-1-1　部分搜索

可以得到整个迷宫地图的资料，这样就可以求出最优路径。但采用这种方式所使用的搜索时间会比较长。

本任务选择方式一进行搜索。

2. 搜索法则

迷宫机器人在搜索迷宫的时候，需要时刻记录自身的迷宫坐标与对应的墙壁资料，漫无目的地搜索会非常影响效率。为了节约搜索时间，在搜索的时候，会采用一些搜索法则来提高搜索效率。

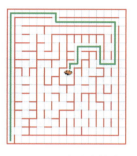

图3-1-2　全迷宫搜索

常用基础搜索法则有三种：右手法则、左手法则和中心法则，如图3-1-3所示。

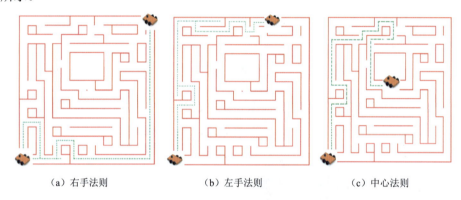

（a）右手法则　　　　　（b）左手法则　　　　　（c）中心法则

图3-1-3　右手法则、左手法则、中心法则

右手法则：当迷宫机器人的前进方向有多个可供选择时，优先向右转，其次直行，最后左转。

左手法则：当迷宫机器人的前进方向有多个可供选择时，优先向左转，其次直行，最后右转。

中心法则：当迷宫机器人的前进方向有多个可供选择时，优先朝向终点的方向转动。

二、迷宫搜索程序控制

接下来实现迷宫机器人的搜索和法则。

首先启动tqd_Maze robot 功能包中的tqd.launch.py。

```
$ cd ~/Maze robot_ws
$ source install/setup.bash
$ ros2 launch tqd_Maze robot  tqd.launch.py
```

● 视　频

迷宫搜索

复制完整程序文件contest.py。

为了实现底层驱动和顶层逻辑的分离，创建class Maze robot，并继承class Drive。class Drive作为驱动类，class Maze robot作为逻辑类。所有的迷宫机器人上层逻辑都在class Maze robot中实现。

下面主要介绍与搜索策略相关的方法，完整的代码请扫描二维码查看。

1. 搜索方法

```python
def mazesearch(self):
    while 1:
        if self.destinationcheck():
        # 如果到了终点，则返回起点，部分搜索
            destination = (self.mouse_x, self.mouse_y)
            self.turnback()
            self.objectgoto(0, 0)
            return (1, destination)
        else:       # 否则一直搜索
            crosswaycount = self.crosswaycheck()
            if crosswaycount:
                if crosswaycount>1:
                    self.crosswaystack.append((self.mouse_x, self.mouse_y))
                    self.crosswaychoice()
                    self.moveoneblock()
                if crosswaycount==1:
                    self.crosswaychoice()
                    self.moveoneblock()
            else:
                self.turnback()
                self.objectgoto(*self.crosswaystack.pop())
```

2. 右手法则

```python
def rightmethod(self):
    frontpath = '1' if self.f_dis > 0.2 else '0'
    rightpath = '1' if self.r_dis > 0.2 else '0'
    backpath = '1'
    leftpath = '1' if self.l_dis > 0.2 else '0'
    if rightpath == '1':            # 右侧有路
        if self.dir == 0:
            flag = self.mapblock[self.mouse_x+1, self.mouse_y][0]
        elif self.dir == 1:
            flag = self.mapblock[self.mouse_x, self.mouse_y-1][0]
        elif self.dir == 2:
            flag = self.mapblock[self.mouse_x-1, self.mouse_y][0]
        elif self.dir == 3:
            flag = self.mapblock[self.mouse_x, self.mouse_y+1][0]
        if flag == '0':
            self.turnright()
```

```python
                    return

            if frontpath == '1':            # 前方有路
                if self.dir == 0:
                    flag = self.mapblock[self.mouse_x, self.mouse_y+1][0]
                elif self.dir == 1:
                    flag = self.mapblock[self.mouse_x+1, self.mouse_y][0]
                elif self.dir == 2:
                    flag = self.mapblock[self.mouse_x, self.mouse_y-1][0]
                elif self.dir == 3:
                    flag = self.mapblock[self.mouse_x-1, self.mouse_y][0]
                if flag == '0':
                    return

            if leftpath == '1':             # 左侧有路
                if self.dir == 0:
                    flag = self.mapblock[self.mouse_x-1, self.mouse_y][0]
                elif self.dir == 1:
                    flag = self.mapblock[self.mouse_x, self.mouse_y+1][0]
                elif self.dir == 2:
                    flag = self.mapblock[self.mouse_x+1, self.mouse_y][0]
                elif self.dir == 3:
                    flag = self.mapblock[self.mouse_x, self.mouse_y-1][0]
                if flag == '0':
                    self.turnleft()
                    return
```

3. 左手法则

```python
    def leftmethod(self):
        frontpath = '1' if self.f_dis > 0.2 else '0'
        rightpath = '1' if self.r_dis > 0.2 else '0'
        backpath = '1'
        leftpath = '1' if self.l_dis > 0.2 else '0'
        if leftpath == '1':             # 左侧有路
            if self.dir == 0:
                flag = self.mapblock[self.mouse_x-1, self.mouse_y][0]
            elif self.dir == 1:
                flag = self.mapblock[self.mouse_x, self.mouse_y+1][0]
            elif self.dir == 2:
                flag = self.mapblock[self.mouse_x+1, self.mouse_y][0]
            elif self.dir == 3:
                flag = self.mapblock[self.mouse_x, self.mouse_y-1][0]
            if flag == '0':
                self.turnleft()
                return

        if frontpath == '1':            # 前方有路
            if self.dir == 0:
                flag = self.mapblock[self.mouse_x, self.mouse_y+1][0]
            elif self.dir == 1:
```

```
                flag = self.mapblock[self.mouse_x+1, self.mouse_y][0]
            elif self.dir == 2:
                flag = self.mapblock[self.mouse_x, self.mouse_y-1][0]
            elif self.dir == 3:
                flag = self.mapblock[self.mouse_x-1, self.mouse_y][0]
            if flag == '0':
                return

        if rightpath == '1':              # 右侧有路
            if self.dir == 0:
                flag = self.mapblock[self.mouse_x+1, self.mouse_y][0]
            elif self.dir == 1:
                flag = self.mapblock[self.mouse_x, self.mouse_y-1][0]
            elif self.dir == 2:
                flag = self.mapblock[self.mouse_x-1, self.mouse_y][0]
            elif self.dir == 3:
                flag = self.mapblock[self.mouse_x, self.mouse_y+1][0]
            if flag == '0':
                self.turnright()
                return
```

4. 中心法则

```
    def centralmethod(self):
        if self.mouse_x < 8:
            if self.mouse_y < 8:        # Maze robot 位于左下角
                if self.dir == 0:
                    self.frontrightmethod()
                elif self.dir == 1:
                    self.frontleftmethod()
                elif self.dir == 2:
                    self.leftmethod()
                elif self.dir == 3:
                    self.rightmethod()
            else:                       # Maze robot 位于左上角
                if self.dir == 0:
                    self.rightmethod()
                elif self.dir == 1:
                    self.frontrightmethod()
                elif self.dir == 2:
                    self.frontleftmethod()
                elif self.dir == 3:
                    self.leftmethod()
        else:
            if self.mouse_y < 8:        # Maze robot 位于右下角
                if self.dir == 0:
                    self.frontleftmethod()
                elif self.dir == 1:
                    self.leftmethod()
                elif self.dir == 2:
                    self.rightmethod()
```

```
            elif self.dir == 3:
                self.frontrightmethod()
        else:                               # Maze robot 位于右上角
            if self.dir == 0:
                self.leftmethod()
            elif self.dir == 1:
                self.rightmethod()
            elif self.dir == 2:
                self.frontrightmethod()
            elif self.dir == 3:
                self.frontleftmethod()
```

任务二　掌握迷宫信息的存储与读取

本任务目的是掌握迷宫信息的存储与读取。

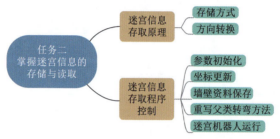

一、迷宫信息存取原理

迷宫机器人搜索迷宫的目的是找到终点，在搜索的过程中需要将迷宫地图信息记录下来，这样才能在第二次行走时直接读取保存的信息，冲刺到终点。

迷宫是由16×16个单元格组成的，很明显建立一个二维数组，x和y分别对应数组的row和col，单元格墙壁资料保存为对应value。迷宫机器人每更新一次坐标，就应当保存一次墙壁资料，直至到达终点。

1. 存储方式

对于一个单元格的墙壁资料，迷宫机器人进来的方向一定是路口（起点除外），另外三面需要根据激光雷达插件检测的数据进行判断，大于0.12 m是路口，小于0.12 m是挡板。路口的有无，正好对应1或0，因此本任务使用一个长度为4的字符串，每一位用1或0表示一个单元格的墙壁资料，例如"0001"，如图3-1-4所示。

① 第一位表示单元格左侧墙壁，

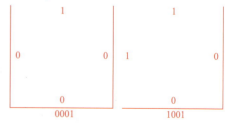

图3-1-4　墙壁资料示意图

1表示有路口，0表示没有路口；

② 第二位表示单元格下方墙壁，1表示有路口，0表示没有路口；

③ 第三位表示单元格右侧墙壁，1表示有路口，0表示没有路口；

④ 第四位表示单元格上方墙壁，1表示有路口，0表示没有路口。

墙壁信息存储如图3-1-5所示。

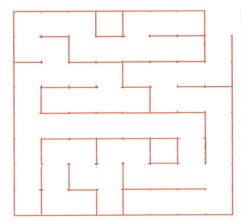

0110	1010	1100	0000	0110	1010	1000	0111
0011	1100	0011	1010	1011	1010	1000	0101
0110	1011	1010	1100	0010	1110	1010	1001
0101	0010	1010	1011	1000	0011	1010	1100
0111	1010	1010	1010	1010	1010	1100	0101
0101	0110	1100	0110	1100	0000	0101	0101
0101	0101	0011	1101	0011	1010	1011	1101
0001	0011	1000	0001	0010	1010	1010	1001

图3-1-5 墙壁信息存储

2. 方向转换

由于迷宫机器人在检测四周墙壁时，是以迷宫机器人自身为参照进行的，也就是相对方向资料。在将资料保存到数组中时，推荐将其转换为绝对方向资料，如图3-1-6所示。

相对方向：左　后　右　前。

绝对方向：left bottom right up。

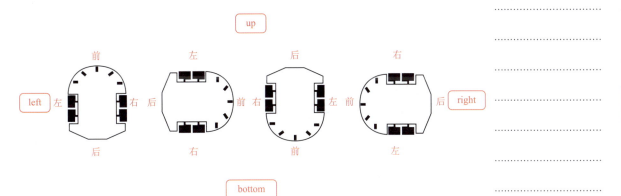

图3-1-6 根据朝向保存资料

二、迷宫信息存取程序控制

首先启动tqd_Maze robot 功能包中的tqd.launch.py。

```
$ cd ~/Maze robot _ws
$ source install/setup.bash
$ ros2 launch tqd_Maze robot  tqd.launch.py
```

复制完整程序文件contest.py。

1. 参数初始化

信息存储与读取视频

```
def valuesinit(self):
    self.mapblock = np.full((16, 16), '00000')
                    # 2D array,墙壁资料初始化, flag left down right up
                    # flag: 是否走过, 0 or 1
                    # left: 是否有路, 0 or 1
                    # down: 是否有路, 0 or 1
                    # right: 是否有路, 0 or 1
                    # up: 是否有路, 0 or 1

    self.mapblock[0,0] = '10001'# 起点墙壁初始化
    self.crosswaystack = []
    self.mouse_x = 0            # x,y起点坐标，这是在迷宫中的坐标
    self.mouse_y = 0
    self.dir = 0                # Maze robot 车头，0:up;
                                  1:right; 2:down; 3:left
```

2. 坐标更新

```
def __coordinateupdate(self):
    if self.dir == 0:           # 车头向上
        self.mouse_y += 1
    elif self.dir == 1:         # 车头向右
        self.mouse_x += 1
    elif self.dir == 2:         # 车头向下
        self.mouse_y -= 1
    elif self.dir == 3:         # 车头向左
        self.mouse_x -= 1
```

3. 墙壁资料保存

```
def __savewallinfo(self):
    """墙壁信息保存，禁止修改"""
    frontpath = '1' if self.f_dis > 0.2 else '0'
    rightpath = '1' if self.r_dis > 0.2 else '0'
    backpath = '1'
    leftpath = '1' if self.l_dis > 0.2 else '0'
    # 墙壁资料赋值
    if self.dir == 0:
        self.mapblock[self.mouse_x, self.mouse_y] = ''.join(('1', leftpath, backpath, rightpath, frontpath))
```

```
            elif self.dir == 1:
                self.mapblock[self.mouse_x, self.mouse_y] = ''.join(('1',
backpath, rightpath, frontpath, leftpath))
            elif self.dir == 2:
                self.mapblock[self.mouse_x, self.mouse_y] = ''.join(('1',
rightpath, frontpath, leftpath, backpath))
            elif self.dir == 3:
                self.mapblock[self.mouse_x, self.mouse_y] = ''.join(('1',
frontpath, leftpath, backpath, rightpath))
```

4. 重写父类转弯方法

迷宫机器人的朝向是在转弯时发生改变的，因此重写父类的turnright()、turnleft()和turnback()方法。

```
    def turnright(self):
        self.dir = (self.dir+1)%4
        return super(Maze_robot , self).turnright()

    def turnleft(self):
        self.dir = (self.dir+3)%4
        return super(Maze_robot , self).turnleft()

    def turnback(self):
        self.dir = (self.dir+2)%4
        return super(Maze_robot , self).turnback()
```

5. 迷宫机器人运行

移动一个单元格，然后更新坐标，保存墙壁信息的功能。

```
    def moveoneblock(self):
        self.move()
        self.__coordinateupdate()
        if self.mapblock[self.mouse_x, self.mouse_y][0]=='0':
            self.__savewallinfo()
```

执行程序后，迷宫机器人会按照单元格移动，每移动一个单元格，更新一次坐标。若墙壁资料数组标志位是'0'，则保存坐标。

任务三　掌握迷宫最优路径规划

本任务目的是掌握最优路径规划的原理，学习如何制作等高图。

一、迷宫最优路径制作原理

假设迷宫机器人已经搜索完整个迷宫或者只搜索了包含起点和终点的部分迷宫,且记录了已走过的每个迷宫单元格的墙壁资料,那么它怎样根据已有信息找出一条从起点到终点最优的路径呢?

下面引入等高图的概念和制作方法。

等高图就是等高线地图的简称,犹如一般地图可以标出同一高度的地区范围,或气象报告时的等气压图,可以标出相等气压的范围及大小,如图3-1-7所示。那么等高图运用在迷宫地图上,可以计算每一个迷宫格与迷宫起点或迷宫终点的距离值。

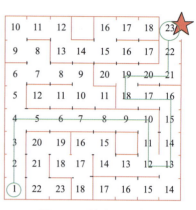

图3-1-7 等高图示意图

迷宫机器人搜索到达终点时,已经保存了所有走过的墙壁资料,其中包含起点和终点,制作以起点为目标点的等高图,制作步骤如下:

① 起点等高值标记为1,相邻且可以到达的单元格等高值+1,以此类推。

② 当单元格有多个方向可以前进时,分别沿着每一个方向独立计算等高值,直至遇到死胡同或交叉路口。

③ 若两条支路存在交叉,在计算等高值时,取较小的等高值标记为该交叉单元格的等高值。

等高值计算完毕后,每个单元格上显示的就是距离目标点(起点)的最小步数,迷宫机器人从终点沿着等高值减小的方向行走,就可以最快回到起点。同时这也就是最短路径。

迷宫机器人从起点再次冲刺到终点是一个逆向的过程,不再解释。

二、迷宫最优路径程序控制

首先启动tqd_Maze robot 功能包中的tqd.launch.py。

```
$ cd ~/Maze_robot_ws
$ source install/setup.bash
$ ros2 launch tqd_Maze_robot tqd.launch.py
```

复制完整程序文件contest.py。

1. 制作等高图

```
def mapstepedit(self, dx, dy):
```

最优路径规划

```
"""制作等高图，禁止修改"""
self.mapstep = np.full((16,16), 255)
step = 1
n = 1
stack = []
stack.append((dx, dy))
cx = dx
cy = dy

while n:
    self.mapstep[cx, cy] = step
    step += 1
    
    count = 0
    # 统计当前坐标有几个可前进的方向
    if (self.mapblock[cx, cy][-1] == '1') and (self.mapstep[cx, cy+1]>step):
        count += 1
    if (self.mapblock[cx, cy][-2] == '1') and (self.mapstep[cx+1, cy]>step):
        count += 1
    if (self.mapblock[cx, cy][-3] == '1') and (self.mapstep[cx, cy-1]>step):
        count += 1
    if (self.mapblock[cx, cy][-4] == '1') and (self.mapstep[cx-1, cy]>step):
        count += 1
    
    if count == 0:
        cx, cy = stack.pop()
        step = self.mapstep[cx, cy]
        n -= 1
    else:
        if count>1:
            stack.append((cx, cy))
            n += 1
        # 随便挑一个方向，不影响结果，因为会全部走一遍
        if (self.mapblock[cx, cy][-1] == '1') and (self.mapstep[cx, cy+1]>step):
            cy += 1
            continue
        if (self.mapblock[cx, cy][-2] == '1') and (self.mapstep[cx+1, cy]>step):
            cx += 1
            continue
        if (self.mapblock[cx, cy][-3] == '1') and (self.mapstep[cx, cy-1]>step):
            cy -= 1
            continue
        if (self.mapblock[cx, cy][-4] == '1') and (self.mapstep[cx-1, cy]>step):
            cx -= 1
```

2. 向目标点运动

```
        def objectgoto(self, dx, dy):
"""向目标点运动，禁止修改"""
            self.mapstepedit(dx, dy)
            temp_dir = 0
            cx = self.mouse_x
            cy = self.mouse_y
            while (cx!=dx) or (cy!=dy):
                # 沿着等高值减小的方向运行，直至到达目标点
                step = self.mapstep[cx, cy]
                if (self.mapblock[cx, cy][-1] == '1') and (self.mapstep[cx, cy+1]<step):
                    temp_dir = 0
                elif (self.mapblock[cx, cy][-2] == '1') and (self.mapstep[cx+1, cy]<step):
                    temp_dir = 1
                elif (self.mapblock[cx, cy][-3] == '1') and (self.mapstep[cx, cy-1]<step):
                    temp_dir = 2
                elif (self.mapblock[cx, cy][-4] == '1') and (self.mapstep[cx-1, cy]<step):
                    temp_dir = 3
                d_dir = (temp_dir - self.dir + 4 )%4
                if d_dir == 1 :
                    self.turnright()
                elif d_dir == 2 :
                    self.turnback()
                elif d_dir == 3 :
                    self.turnleft()
                self.moveoneblock()
                cx = self.mouse_x
                cy = self.mouse_y
```

当迷宫机器人到达终点或进入死胡同的时候，会调用objectgoto()返回上一个岔路口。基于等高图的行走，可以极大地降低运行时间。

思考与总结

1. 迷宫机器人的常用搜索策略有哪些？

2. 在制作最优路径时还有没有提高空间？

3. 迷宫机器人在运行时涉及相对方向与绝对方向的转换，检测的墙壁信息是相对方向的，需要转为绝对方向后再存储；在读取已有墙壁信息时，需要将绝对方向转回相对方向再进行控制。

项目二

迷宫机器人竞赛虚拟仿真实训

 学习目标

1. 了解虚拟仿真评测系统的使用方式。
2. 了解使用虚拟仿真评测系统进行一场比赛。

任务一 学习虚拟仿真评测系统

本任务目的是学习使用TQD-OC迷宫机器人虚拟仿真评测系统。

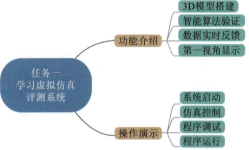

一、功能介绍

TQD-OC V2.0是天津启诚伟业科技有限公司根据虚拟仿真教育教学需求,开发设计的新一代迷宫机器人虚拟仿真评测系统,如图3-2-1所示。

本系统部署了3D模型搭建、多传感器融合、智能算法验证、数据实时反馈等功能,如图3-2-2所示;提供第一视角显示,体验身临其境的感觉;支持左手算法、右手算法、中心算法和洪水算法,可以轻松实现标准虚拟迷宫的搜索和最短路径的优化。

在仿真运行过程中,迷宫机器人的速度、偏移

图3-2-1 TQD-OC迷宫机器人虚拟仿真竞赛评测系统

量、转弯角度等数据实时反馈，方便参赛者进行数据分析。本系统还提供成绩记录的功能，系统

| 3D模型搭建 | 多传感器融合 | 智能算法验证 | 数据实时反馈 |

图3-2-2　评测系统特点

实时显示迷宫时间和运行时间，记录最优成绩，并在竞赛结束后自动排序。系统支持模式切换，在调试模式下，将启用控制接口，参赛者可以通过校正参数使迷宫机器人的运行更加高效。虚拟仿真评测系统部署了上传和下载的功能，一键切换中英文，满足国际选手与国内选手使用同一平台实时竞技的需求。同时还提供大量不同难度的3D迷宫与迷宫机器人模型，根据竞赛需求自由切换。

1. 3D模型搭建

评测系统融合3D模型一键搭建的功能，可以将TQD迷宫模型文件，一键生成3D脚本文件，方便用户自定义模型，如图3-2-3所示。

2. 智能算法验证

评测系统提供智能算法验证的功能，如图3-2-4所示，以平面的显示方式，快速生成算法运行轨迹，辅助用户进行算法设计。

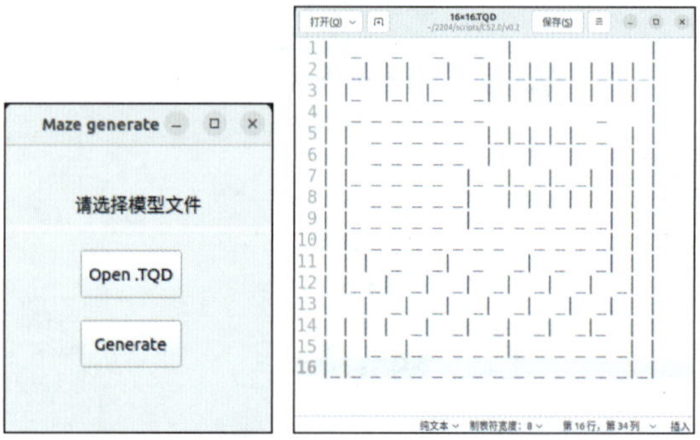

图3-2-3　3D模型搭建 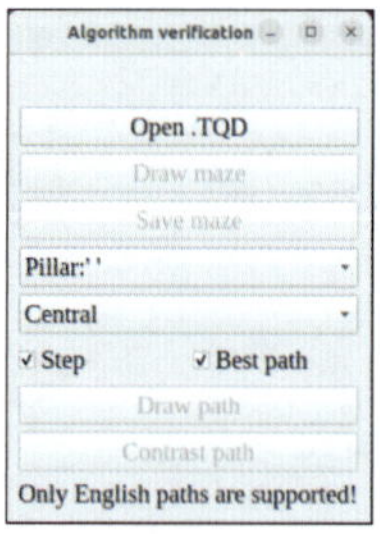　　　　　图3-2-4　智能算法验证

智能算法验证工具有以下功能：

① Open .TQD：加载一个.TQD迷宫模型文件。

② Draw maze：绘制迷宫，将模型文件中的迷宫绘制成图片。

③ Save maze：保存图片，将绘制的迷宫、各种算法的运行轨迹等都保存到本地图片中。

④ Pillar：.TQD文件中对于立柱的描述方式，默认空格，可更改为'|'。

⑤ Central：自由选择算法，目前支持左手算法、右手算法和中心算法。

⑥ Step：是否显示等高值，起点为1。

⑦ Best path：是否绘制最优路径。

⑧ Draw path：绘制机器人使用当前选择的算法时的运行轨迹。

⑨ Contrast path：对比迷宫原图（Original）、中心（Central，绿色）、右手（Right，黄色）和左手（Left，蓝色），如图3-2-5所示。

3. 数据实时反馈

评测系统提供数据实时反馈的功能，迷宫机器人在运行过程中，所有数据均动态可视化，方便用户进行数据分析，如图3-2-6所示。

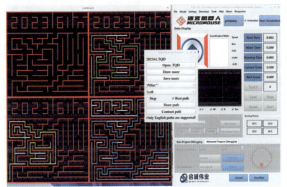
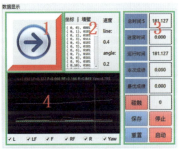

图3-2-5　迷宫轨迹对比　　　　图3-2-6　数据显示

标号1：朝向和墙壁显示。

箭头表示迷宫机器人的车头朝向，四条框线表示当前机器人四周的墙壁信息，红色表示有墙，白色表示无墙。车头朝向是算法程序设计的重要元素之一，辅助参赛者掌握迷宫机器人的转向变化。墙壁有无的动态显示，辅助参赛者了解迷宫机器人检测是否正确。

标号2：坐标和墙壁显示。

坐标列显示当前迷宫机器人在迷宫中的位置。墙壁资料列，四个数字分别对应左、下、右和上，四个方向的墙壁有无，1表示该方向有路，0表示该方向没有路。直线速度、旋转速度也同步显示。墙壁资料可以作为图中显示的验证。坐标的实时显示，辅助参赛者验证迷宫机器人是否发生坐标计算错误。

标号3：计分系统显示。

记录迷宫机器人在迷宫中的各项时间数据。单击"启动"按钮启动程序，迷宫机器人将开始运行；单击"重置"按钮，将终止迷宫机器人的运行。

总时间：指迷宫机器人在迷宫中的运行总时间。

迷宫时间：指迷宫机器人从第一次启动到每次运行开始的那段时间。

运行时间：指迷宫机器人每次运行的时间。

本次成绩：指本次运行取得的成绩。

最优成绩：指迷宫机器人在多次运行中取得的最优成绩。

碰触：指迷宫机器人发生错误，需要复位回到起点重新启动的次数。

本次成绩 = 迷宫时间/30 + 运行时间 +（触碰-1）×5

式中，除以30是为了降低搜索耗时所占的比重；乘以5是为了奖励无碰触，惩罚多次触碰。用户可以在计分规则中进行自定义设置。

标号4：传感器数据显示。

迷宫机器人的激光传感器和偏航角数据都会直观地从绘制的波形中显示出来。辅助参赛者进行参数调试，了解迷宫机器人的偏移情况。

4. 第一视角显示

评测系统提供第一视角显示的功能，仿佛置身于迷宫之中，体验身临其境的感觉，如图3-2-7所示。

图3-2-7　第一视角显示

二、操作演示

1. 系统启动

视　频

TQD-OC V2.0
评测系统演示

在登录页面输入正确的账号和密码，即可跳转到主页面，如图3-2-8所示。

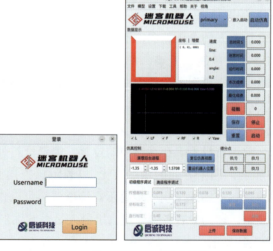

图3-2-8　登录页面和主页面

2. 仿真控制

选择合适的难度，单击"启动仿真"按钮即可打开仿真视图，如图3-2-9所示。

当前选择使用的迷宫和迷宫机器人可以在模型菜单中进行更改。

评测系统提供两种仿真模式：分离式和嵌入式。

分离式：仿真视图独立启动，占用系统资源较多。

嵌入式：选中"嵌入启动"单选按钮，仿真视图将嵌入评测系统框架中，以平面形式启动，有效节省系统资源，推荐选择嵌入式启动方式。

仿真视图打开后，就可以使用仿真控制中的相关功能对仿真环境和迷宫机器人进行控制，如图3-2-10所示。

清理后台进程：包括Gazebo进程、debug脚本进程、contest脚本进程、camera脚本进程等。

复位仿真视图：对仿真视图进行复位，包括迷宫姿态和迷宫机器人姿态。

重设机器人位置：重新推送迷宫机器人，在新的位置并以新的偏航角出现在仿真视图中。

图3-2-9　打开仿真视图

图3-2-10　仿真控制

3. 程序调试

评测系统提供多个调试接口，在设置中切换调试模式后，即可使用相关功能。

（1）初级程序调试

传感器标定：设置传感器检测的相关阈值。

坐标标定：设置单位单元格的大小。

直行标定：设置直行的速度和直行校正车姿的校正系数，如图3-2-11所示。

（2）高级程序调试

方向校准：校准方向与偏航角的对应关系。

转弯验证：控制迷宫机器人转弯，验证转弯角度是否准确，如图3-2-12所示。

图3-2-11　初级程序调试

图3-2-12　高级程序调试

4. 程序运行

程序调试完毕后，在设置中切换为竞赛模式，就可以运行竞赛程序了，如图3-2-13所示。

单击"启动"按钮，将开始运行contest.py，UI中同步显示各项运行数据。

① 若迷宫机器人发生了错误，无法继续运行，可以单击"碰触"按钮，终止contest.py进程。

② 再单击仿真控制中的"复位仿真视图"按钮，恢复迷宫和

图3-2-13　计分系统

机器人姿态。

③ 再重新单击"启动"按钮,控制迷宫机器人重新运行。

④ 单击"停止"按钮,结束本次竞赛。

⑤ 单击"保存"按钮,保存本次竞赛数据。

⑥ 单击"复位"按钮,结束本次竞赛并清空所有数据。

任务二 了解虚拟仿真竞赛

视频
虚拟竞赛

本任务目的是学习使用TQD-OC迷宫机器人虚拟仿真评测系统进行一场竞赛。

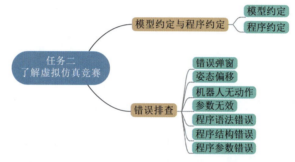

一、模型约定与程序约定

虚拟仿真竞赛是使用代码来完成的迷宫机器人竞赛,参赛者对代码拥有100%的自我创作能力,因此为了保证比赛公平公正的顺利进行,需要所有参赛者遵守一定的规则。

1. 模型约定

虚拟仿真竞赛中,所有参赛者使用相同的3D迷宫进行竞技。迷宫模型由竞赛组委会提供,因此不需要特别说明。迷宫机器人模型可以使用组委会提供的模型,也可以自行制作,但需遵守以下约定:

① 迷宫机器人的最高速度不得超过0.4 m/s。

② 迷宫机器人的长和宽不超过0.2 m×0.2 m,高度不得超过0.05 m。

③ 迷宫机器人仅限使用两轮差速插件和激光雷达插件。

④ 迷宫机器人在迷宫中的初始位置,请根据组委会下发的迷宫,自行设置。

2. 程序约定

竞赛程序文件需使用Python编写,命名为contest.py。在编写程序的时候同样需要遵守以下约定。

① 保留主程序入口,否则无法启动程序。

② 每当虚拟迷宫机器人坐标发生改变时，在指定话题上发布消息：

```
话题名称: "transmit",
消息类型: std_msgs.msg.String
消息内容: 当前虚拟迷宫机器人的各项参数（改变后的），是一个以逗号','分隔的
字符串。
"l_dis, f_dis, r_dis, speed_x, speed_z, dir, mouse_x, mouse_y,
mapblock[mouse_x, mouse_y]"
    l_dis                       :水平左侧挡板距离
    f_dis                       :正前方挡板距离
    r_dis                       :水平右侧挡板距离
    speed_x                     :平移速度
    speed_z                     :旋转速度
    dir                         :车头方向
    mouse_x                     :x坐标
    mouse_y                     :y坐标
    mapblock[mouse_x, mouse_y]: 四周墙壁，用'0'和'1'表示是否有路，长度为
4，顺序为左下右上
```

③ 不得使用图像处理程序，一经发现，将取消比赛资格。

二、错误排查

在运行评测系统或运行竞赛程序时，若出现了错误，可参考以下示例进行错误排查。

1. 错误弹窗

若运行时出现了错误弹窗，请根据弹窗提示进行合理修改，如图3-2-14所示。

图3-2-14 错误弹窗

2. 姿态偏移

若迷宫机器人在起始位置发生了姿态偏移，建议先复位迷宫和迷宫机器人，否则会影响后续的运行。单击评测系统面板上的"复位仿真视图"按钮，复位迷宫和迷宫机器人。如果是分离式启动，还可以选择Edit菜单中的Reset World命令，或使用组合键【Ctrl+R】，复位迷宫和迷宫机器人，如图3-2-15所示。

3. 机器人无动作

若单击"启动"按钮，迷宫机器人无动作，请单击"清理后台进程"按钮，或重新启动评测系统。

图3-2-15 复位仿真视图

4. 参数无效

若参数校准后，在竞速模式下没有生效，说明没有保存，只有单击"发布数据"按钮后，校准过的参数才会保存，运行contest.py时才会生效。

5. 程序语法错误

编写Python程序时，建议使用VScode或Pycharm，可以提示绝大部分语法错误，如图3-2-16～图3-2-19所示。

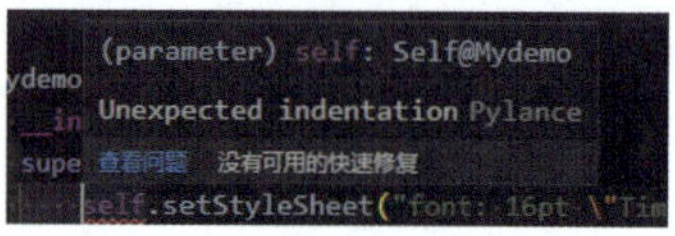

图3-2-16　缩进错误

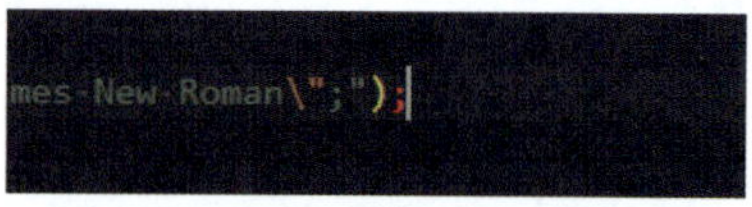

图3-2-17　多余分号

图3-2-18　缺少冒号

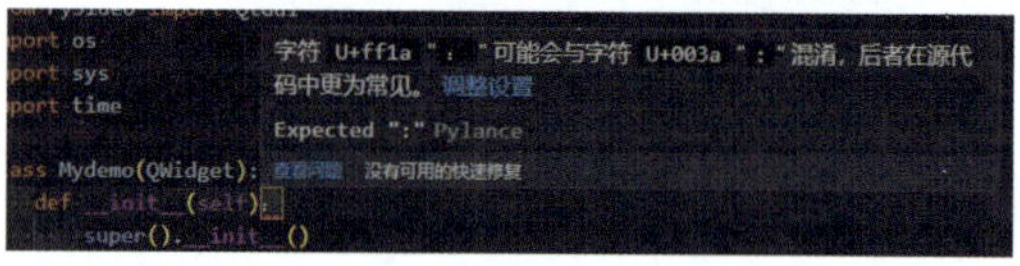

图3-2-19　中英文混乱

6. 程序结构错误

（1）while语句

在例程中主要用于进入一个状态，比如直行状态或旋转状态。

（2）if语句

在例程中主要用于条件判断，满足条件执行什么，比如结束直行状态或旋转状态，墙壁有无的判断等。

（3）嵌套控制语句

在对复杂情况进行判断时会使用嵌套的控制语句，这时一定要注意层次结构。排查这种错误是非常困难的。

7. 程序参数错误

需要根据实际的运行效果来判断，比如：

①单位单元格大小错误：导致所有的坐标错误，这是非常严重的。

②传感器阈值不准确：导致无法在中心线上行走，或者转弯错误，墙壁有无判断错误等。

③转弯角度有偏差：导致姿态偏差，角度过大、过小，会影响后续运行。

④直线速度或旋转速度过快：速度过快会产生一些不可预料的结果，影响后续的动作，比如直接撞墙、原地转圈等。

思考与总结

1. 如何切换虚拟仿真评测系统的工作模式？

2. 为了使迷宫机器人可以稳定运行，在竞赛前一定要先调试它的运行参数。

附录

附录A

2022年首届世界职业院校技能大赛——迷宫机器人赛项规则

一、竞赛形式

本赛项采用线上的形式开展比赛。

各个参赛队在本校或经批准的其他学校内设置符合要求的赛场，按规定时间统一开赛，赛场内架设摄像机360度无死角拍摄比赛现场。比赛采用线上方式进行检录抽签等活动。现场裁判和监督仲裁人员各1名入校执裁和监督，各省厅派纪检人员1名参与。比赛结束后，各参赛队在规定时间内提交比赛结果和相关赛场监控视频。评分裁判集中进行评分，仲裁全程线上监督各队比赛。

二、竞赛内容

竞赛共3个模块，包括模块A：虚拟仿真环节；模块B：迷宫竞速环节；模块C：展示答辩环节。

竞赛环节	模块 A	模块 B	模块 C
	虚拟仿真	迷宫竞速	展示答辩
比赛时长	约3 h	约2 h	约2 h
工匠精神	考察职业素养与安全意识，共10分，为扣分项		

在比赛开始前，由裁判长在线上发布"竞赛口令"，工作人员将抽签编号、竞赛流程等相关信息张贴在赛场醒目位置。比赛时，按A、B、C三个模块顺序进行。选手在竞赛场地内使用大赛规定设备进行操作，每个模块比赛结束后按照裁判指令开始下一模块的比赛。整个比赛完成后，选手在现场裁判的监督下，将录像视频文件在规定时间内统一上传到大赛指定平台，供裁判评分。

摄像头和录像机拍摄的画面中须全程展现正确的"竞赛口令"。

三、竞赛时间流程

按照大赛组委会规定的时间,8时整开始检录,8时30分开始比赛,竞赛时长420分钟。

主要包括报到检录、虚拟仿真(约3 h)、迷宫竞速(约2 h)、展示答辩(约2 h)和成绩汇总等环节,具体安排如下:

日 期	时 间	工作事项	参加人员	方 式
竞赛前1日	10:00—12:00	裁判工作会议	裁判长、裁判员、监督仲裁组	线上线下结合
	13:00—14:00	领队会	各参赛队领队、裁判长	线上
	14:00—16:00	赛场调试	各参赛队领队、现场裁判、监督仲裁员	线上线下结合
	16:00	检查封闭赛场	裁判长、监督仲裁员	线上
竞赛日	8:00—8:30	选手检录和抽签等	参赛选手、加密裁判、保障组	线上线下结合
	8:30—11:30	模块A 虚拟仿真	参赛选手、现场裁判、裁判长、保障组	线上
	11:30—13:00	午餐		
	13:00—15:00	模块B 迷宫竞速	参赛选手、现场裁判、裁判长、保障组	线上线下结合
	15:00—17:00	模块C 展示答辩	参赛选手、现场裁判、裁判长、保障组	线上
	17:00—19:00	成绩汇总和审核	裁判、监督仲裁组	线上线下结合

四、成绩评审

1. 裁判需求表

裁判和监督人员采取"集中-分散"相结合的工作方式。大赛共设置裁判长1名、仲裁1名、评分裁判6名、现场裁判11名、现场监督仲裁11名、现场纪检人员11名。

其中,裁判长、仲裁、评分裁判原则上集中线下办公,11个参赛校比赛现场各派现场裁判1名、监督仲裁人员1名、纪检人员1名。

现场裁判和现场监督仲裁:由世校赛执委会在本省/市抽取产生(与参赛队所在学校有直接关系的应回避),负责竞赛现场全程裁决和监督工作。

评分裁判:由世校赛执委会在大赛裁判库中抽取,承办省市裁判比例不低于50%。

现场纪检人员:由参赛校所在省厅派遣。

竞赛结束后,评分裁判背靠背打分(含机评系统自动打分),专家组和评分裁判原则上集中到承办校集体办公。

序　　号	裁判类别	需求人数	工作场所
1	裁判长	1	主办校
2	仲裁	1	主办校
3	评分裁判	6	主办校
4	现场裁判	11	各参赛校
5	现场监督	11	各参赛校
6	纪检人员	11	各参赛校

2. 评审方式

采用线下与线上评审相结合的方式对模块A、B、C逐个进行评分。

① 比赛结束后立即开始评分。评分时各参赛队选手必须离开工位，不得再进行任何操作。技术支持人员在赛场内等候评分裁判指令。竞赛设备保持比赛结束选手离场时的状态。

② A模块评分方法：评分裁判和现场裁判紧密配合，远程监控工位摄像头，观察选手操作过程是否规范，结合虚拟仿真运行结果进行评分。监督仲裁员全程监督。

③ B模块评分方法：参赛选手按照裁判指令进行现场竞速比赛，技术支持人员按裁判要求调整监控视角，保证比赛无死角观测到比赛的全过程。监督仲裁员全程监督。

④ C模块评分方法：参赛选手按照裁判指令进行展示和答辩，评分裁判员依据评分标准进行评分。

五、赛事安排

1. 赛前准备

竞赛前7天，各参赛校选定参赛场地并上报世校赛执委会，完成人员调配、设备调试和环境布置等准备工作。

参与人员：参赛校、保障组、联络员、领队。

2. 赛项说明会

竞赛前15天召开赛项说明会，公布竞赛时间、竞赛方式、环境要求、竞赛流程、注意事项等内容。

参与人员：专家组、裁判长、领队、指导教师。

3. 赛场验收

竞赛前1天，参赛校竞赛环境测试。世校赛执委会专家组、裁判组、监督仲裁组通过腾讯会议（会议号通过参赛校联络员下发，并向保障组短信确认）进行检查验收并测试。验收通过后，赛场封闭贴封条，录制封场视频。

参与人员：参赛校领队及联络员、专家组、裁判组、监督仲裁组、保障组。

4. 进场准备

竞赛当天规定时间前，各参赛校及相关人员进入竞赛场地，保障组工作人员创建本评审组视频会议，用短信通知本评审组参赛队联络员视频会议号。参赛队联络员回复指定手机号码确认："×××（学校名称）参赛队已收到迷宫机器人技术应用赛项腾讯会议号：××× ××× ×××，×月××日上午/下午×时前，做好一切准备。特此确认"。在现场裁判的监督下开封赛场并录制视频，通过视频会议进入相应评审组并调试好所有设备。场内除了参赛选手，现场裁判，现场监督，合作企业技术支持工程师，视频拍摄、转换、上传技术人员和视频连线技术人员之外，不得有其他人员在场；始终保持视频连线，并能全程监视决赛场所。参赛队按时用视频连线计算机登录视频会议，将成员名改为赛位号+工位号。开启外接广角摄像头（一直到竞赛全部事宜结束），由保障组工作人员、现场裁判、监督仲裁组人员等检查场所、场内人员。

5. 身份核验

竞赛当天8时，每个参赛队在规定时间内，通过视频会议与保障组工作人员单独连线，各参赛选手听从保障组工作人员的指挥，逐一在广角摄像头前展示人脸及本人身份证（护照）、学生证、指导教师工作证，保障组工作人员将截屏留存，完成参赛选手的身份核验。

6. 抽定赛位号

加密裁判按参赛队联络员姓氏笔画为序，在监督仲裁组的监督下，抽签决定参赛队的赛位号；每个参赛队使用赛位号进入竞赛专用腾讯会议。参赛团队负责人回复短信确认。

7. 实时录制

由保障组工作人员在统一的时间点连线公布"竞赛特定标识"，由各参赛校固定张贴（或书写）在视频录制始终可见位置。

8. 竞赛答题结果上传

竞赛结束后，参赛选手按题目要求将竞赛答题结果在规定时间内上传至指定地址。

9. 现场录像上传

将录制好的视频文件分别以"W12+迷宫机器人赛项+时段号（封闭赛场、

开封赛场）+赛位号.mp4"命名，比赛现场视频按照下表要求进行命名，采用MP4格式封装，不允许另行剪辑及配音，视频录制软件不限，采用H.264/AVC（MPEG-4 Part10）编码格式压缩；动态码流的码率不低于1 024 kbit/s；分辨率设定为720×576（标清4∶3拍摄）或1 280×720（高清16∶9拍摄）；采用逐行扫描（帧率25帧/秒）。音频采用AAC（MPEG4 Part3）格式压缩；采样率48 kHz；码流128 kbit/s（恒定）。及时将视频上传至赛项指定邮箱（承办校负责）。比赛现场监控视频上传截止时间为比赛结束后60 min内；封闭和开封赛场录制视频上传为竞赛日当天12点前。

模 块	类 别	命 名 格 式
A模块	录屏	W12+迷宫机器人赛项(Maze robot)+A模块(Module A)录屏+工位号.MP4
	工位录像	W12+迷宫机器人赛项+A模块工位后方+工位号.MP4
	全景录像	W12+迷宫机器人赛项+A模块全景+工位号.MP4
B模块	移动录像	W12+迷宫机器人赛项+B模块移动+赛位号(match number).MP4
	全景录像	W12+迷宫机器人赛项+B模块全景+赛位号.MP4
	评分系统录屏	W12+迷宫机器人赛项+B模块评分系统+赛位号.MP4

10. 完成竞赛

各参赛队在完成竞赛全部事宜，并确认视频上传无误后，参赛队负责人回复指定手机号码确认："迷宫机器人赛项×××（赛位号）参赛队已经完成竞赛，特此确认"。

11. 评审

根据竞赛阶段流程要求，评分裁判依据评分标准打分（含机评-系统自动评分）。评审成绩由裁判长统计汇总。

12. 成绩计算及公示

根据既定规则确定最终成绩，成绩评定方法依照赛项规程，并由监督仲裁组进行成绩复核。成绩按组委会统一要求时间公示。

附录B

"迷宫机器人仿真与设计"国际实训课程标准

课程名称：迷宫机器人仿真与设计。

总学时：48；理论学时：16；实践学时：32。

课程类别：专业选修课。

先修课程：无限制。

面向专业：电子与信息大类、装备制造大类。

一、课程定位与课程目标

1. 课程定位

本课程以迷宫机器人作为主要对象，结合高等院校、职业院校、普通院校创新实践课程改革，强调自主创意、动手实践、机电控制、传感等多项技术综合，具有一定的先进性、启发性和实用性。

本课程目的是以迷宫机器人为载体，开展实践教学活动，帮助学生更好地加强动手能力，让学生熟练掌握各种机构的使用方法和组装方法，实现机器人完成预定编程动作，引导学生开发和创造更合理的机构。能够针对机电工程领域的复杂工程问题提出解决方案，熟悉机电系统设计规程，设计满足特定需要的机电产品和机电系统，在机械结构设计、电路设计、软件编程等方面体现创新意识。

2. 课程目标

通过实验了解机电一体化设备的组成和运行，学会分析设计机电系统的结构、性能，提高动手能力和创新技能，对创新和创造技法有一定的了解和应用能力。

（1）知识目标

① 了解机器人基本的机械结构和特点；了解电、气动控制工业技术的基本概念、分类及特点；了解电、气动元件的工作原理、种类和实际应用方式，电、气动控制系统的组成和结构；了解基本的电、气动控制技术的应用；了解现代自动化控制系统的结构和特点。

② 掌握嵌入式微控制器、传感器、智能控制算法、3D虚拟仿真、人工智能等关键技术、核心知识和核心技能。

（2）能力目标

① 掌握机器人虚拟仿真技术，对机器人建模（URDF）、创建仿真环境（Gazebo）以及感知环境(Rviz)等系统性进行仿真。

② 掌握Python编程语言学习与实践。

③ 基于迷宫机器人虚拟仿真系统，实现迷宫机器人在虚拟平台下的构建、通信以及相关算法的实现。

（3）思政教育目标

① 由辩证唯物主义中的认识论，阐述如何发现科学问题，使学生在逻辑思维能力、抽象思维能力以及分析问题与解决问题的能力方面受到初步训练；培养学生辩证唯物主义世界观和科学思维方法。

② 阐述我国科学家在控制领域中的贡献以及治学风范，增强文化自信和民族自豪感，宣扬工匠精神和社会主义核心价值观，引导学生做社会主义核心价值观的坚定信仰者、积极传播者、模范践行者。

③ 机器人控制领域人才紧缺，以及中国对先进技术的迫切需求现状，增强学生对国家民族自强发展的紧迫感，激发和增强学生的民族自尊心和责任感。

二、课程目标实践途径与方法

序号	实践内容	内容提要	学时分配	教学方式
1	迷宫机器人渊源与发展	了解机器人的发展历史、用途、类型	2	讲课、示范、讨论
2	结构基础知识和控制	了解结构的基本单元或构件的结构、功能和用途；了解几种常用结构的工作原理和搭建过程；了解常用控制的类型、特点、功能和用途；了解几种常用结构的控制原理和过程	2	讲课、示范、讨论
3	迷宫机器人虚拟仿真系统、机器人ROS系统、Python语言编程	学习迷宫建模、机器人姿态控制、迷宫搜索算法等相关知识	18	讲课、示范、讨论、实践
4	创意模型的设计	实现迷宫机器人的设计及实践。在设计过程中体会各个环节都要认真、细致、严谨、耐心、坚持，培养学生分析问题、解决问题的能力、团队协作能力及勇于探索、勇于创新、追求真理的科学精神	24	实践

续表

序 号	实践内容	内容提要	学时分配	教学方式
5	总结	检查总结完成的结构模型及功能，制作PPT，讲述制作创新点，心得体会。加强沟通、表达能力的培养	2	检查、总结

三、考核方式及成绩评定

1. 考核类别

考查。

2. 考核方式

考核内容	考核方式	评定标准（依据）	占总成绩比例
过程考核	含到课率、课堂讨论发言、平时作业等	在项目制作过程中发挥的作用及其工作态度和出勤情况	40%
期末考核	实物制作+报告	构思的创新性，结构的合理性，整体效果的可视性，实验报告的科学性和规范性	60%

四、推荐教材与主要参考书

1. 自编讲义
2. 主要参考书

（1）《智能鼠原理与制作》（王超，高艺，宋立红编著，2019年中国铁道出版社出版）。

（2）《机器人仿真与编程技术》（杨辰光，李智军，许扬编著，2018年清华大学出版社出版）。

（3）《机器人系统建模与仿真》（李艳生，杨美美，魏博，等编著，2020年北京邮电大学出版社出版）。

附录C 教学内容与课时安排

序　号	内　容	学　时
第〇篇　迷宫机器人的渊源与发展	项目一　迷宫机器人的发展渊源 项目二　迷宫机器人与世校赛的渊源	理论学时：2
第一篇　迷宫机器人仿真基础	项目一　Python开发入门 项目二　学习环境搭建 项目三　ROS2入门 项目四　Gazebo入门	理论学时：2 实训学时：8
第二篇　迷宫机器人仿真设计	项目一　学习迷宫机器人制作 项目二　了解迷宫机器人基础功能 项目三　掌握迷宫机器人智能控制	理论学时：6 实训学时：12
第三篇　迷宫机器人仿真实践	项目一　学习迷宫机器人智能算法 项目二　迷宫机器人竞赛虚拟仿真实训	理论学时：6 实训学时：12
总计		总学时：48 理论学时：16 实践学时：32

附录D 专业词汇中英对照表

1. 迷宫机器人相关

中文	英文
核心控制模块	main control module
主控芯片	main control chip
输入模块	input module
输出模块	output module
核心板电路	main control circuit
电源电路	power circuit
控制电路	control circuit
外围电路	peripheral circuit
键盘显示电路	keyboard-display circuit
JTAG接口电路	JTAG interface circuit
按键电路	key-pressing circuit
数据传输	data transmission
人机交互系统	human-computer interaction system
吸地风扇技术	suction fan technology
占空比	duty cycle
角速度	angular velocity
红外传感器	IR sensor
红外检测电路	infrared detection circuit
红外线	infrared light
红外校准	infrared calibration
红外强度	infrared intensity
红外发射头	infrared transmitter
红外接收头	infrared receiver
红外PWM发生器模块	PWM signal generator driver module
右前方、左前方、左方、前方、右方	the front-right, the front-left, the left, the front and the right
g段	the g segment

续表

中文	英文
空心杯直流电机	coreless DC motor
步进电动机	stepping motor
电动机驱动电路	motor drive circuit
真值表	truth table
H桥电路	H-bridge circuit
转动（步进电机）	rotate
电子元器件	electronic components
晶振	crystal oscillator
电容	capacitance
限流可调电阻	adjustable current-limiting resistance
数码管	digitron
外围器件	peripheral device
脉冲振荡电路	pulse oscillation circuit
脉冲信号	pulse signal
方波	square wave
感知系统	perceptual system
载波频率	carrier frequency
原理图	schematic diagram
软件界面	software interface
驱动库	driver library

2. ROS2相关

中文	英文
机器人操作系统	robot operating system
工作空间	workspace
功能包	package
构建	build
环境变量	environment variable
节点	node
话题	topic
服务	service
动作	action
计时器	timer

续表

中文	英文
接口	interface
发布	publish
订阅	subscribe
坐标转换	coordinate transformation
四元数	Quaternion
欧拉角	Euler angle
模拟器	simulator

3. Gazebo相关

中文	英文
统一机器人描述格式	unified robot description format
仿真	simulation
视图	view
插件	plugin
高斯噪声	Gaussian noise
差速驱动	differential drive
激光雷达	laser radar
里程计	odometer
偏航角	yaw
动态链接库	dynamic linked library
线速度	linear velocity
角速度	angular velocity
样本	sample

4. 竞赛相关

中文	英文
单元格	cell
墙壁	wall
立柱	post
竞赛场地	competition maze
起点	the start
目的地/终点	the destination
迷宫坐标	the coordinate in the maze

续表

中文	英文
路口	crossing
电子自动计分系统	electronic automatic scoring system
参赛队员	competitor
迷宫机器人竞速比赛	maze robot competition
最优路径	the optimal path
轨迹	trajectory
通道	passage way

5. 智能算法相关

中文	英文
底层驱动程序	the bottom driver program
顶层算法	the top algorithm program
算法	algorithm
策略	strategy
法则（左、右手法则）	rule
右手、左手、中心法则	the right-hand rule, the left-hand rule, the central rule
90°、180°转弯	90-degree turning/180-degree turning
编程并实现	programming and realizing
等高图	step map
循环检测	cycle detection
实现避障	obstacle avoidance
运动姿态的控制	motion attitude control
两轮差速	two-wheel difference speed
路径规划和决策算法	path planning and decision algorithm
结构体	struct
差速控制	differential-speed control
直线运动	straight movement
转弯	turning
校正车姿	correct the attitude
运行校正	attitude correction
核心函数	core function
（驱动步进电机的）时序状态	time sequence status
前进一格	moving forward one cell

续表

中文	英文
按键等待	waiting for button press
判断车姿	determining the attitude
暂停一步	waiting one step
精确转弯控制	accurate turning control
闭环控制	closed-loop control
绝对方向	absolute direction
相对方向	relative direction

5. Related to intelligence algorithm

Chinese	English
底层驱动程序	the bottom driver program
顶层算法	the top algorithm program
算法	algorithm
策略	strategy
法则(左、右手法则)	rule
右手、左手、中心法则	the right-hand rule, the left-hand rule, the central rule
90°、180° 转弯	90-degree turning/180-degree turning
编程并实现	programming and realizing
等高图	step map
循环检测	cycle detection
实现避障	obstacle avoidance
运动姿态的控制	motion attitude control
两轮差速	two-wheel difference speed
路径规划和决策算法	path planning and decision algorithm
结构体	struct
差速控制	differential-speed control
直线运动	straight movement
转弯	turning
校正车姿	correct the attitude
运行校正	attitude correction
核心函数	core function
(驱动步进电机的) 时序状态	time sequence status
前进一格	moving forward one cell
按键等待	waiting for button press
判断车姿	determining the attitude
暂停一步	waiting one step
精确转弯控制	accurate turning control
闭环控制	closed-loop control
绝对方向	absolute direction
相对方向	relative direction

Continued

Chinese	English
四元数	quaternion
欧拉角	Euler angle
模拟器	simulator

3. Related to Gazebo

Chinese	English
统一机器人描述格式	unified robot description format
仿真	simulation
视图	view
插件	plugin
高斯噪声	Gaussian noise
差速驱动	differential drive
激光雷达	laser radar
里程计	odometer
偏航角	yaw
动态链接库	dynamic linked library
线速度	linear velocity
角速度	angular velocity
样本	sample

4. Related to Competition

Chinese	English
单元格	cell
墙壁	wall
立柱	post
竞赛场地	competition maze
起点	the start
目的地/终点	the destination
迷宫坐标	the coordinate in the maze
路口	crossing
电子自动计分系统	electronic automatic scoring system
参赛队员	competitor
迷宫机器人竞速比赛	maze robot competition
最优路径	the optimal path
轨迹	trajectory
通道	passage way

Continued

Chinese	English
步进电动机	stepping motor
电动机驱动电路	motor drive circuit
真值表	truth table
H桥电路	H-bridge circuit
转动(步进电机)	rotate
电子元器件	electronic components
晶振	crystal oscillator
电容	capacitance
限流可调电阻	adjustable current-limiting resistance
数码管	digitron
外围器件	peripheral device
脉冲振荡电路	pulse oscillation circuit
脉冲信号	pulse signal
方波	square wave
感知系统	perceptual system
载波频率	carrier frequency
原理图	schematic diagram
软件界面	software interface
驱动库	driver library

2. Related to ROS2

Chinese	English
机器人操作系统	robot operating system
工作空间	workspace
功能包	package
构建	build
环境变量	environment variable
节点	node
话题	topic
服务	service
动作	action
计时器	timer
接口	interface
发布	publish
订阅	subscribe
坐标转换	coordinate transformation

Appendix D

Chinese-English Comparison Table of Professional Vocabulary

1. Related to maze robot

Chinese	English
核心控制模块	main control module
主控芯片	main control chip
输入模块	input module
输出模块	output module
核心板电路	main control circuit
电源电路	power circuit
控制电路	control circuit
外围电路	peripheral circuit
键盘显示电路	keyboard-display circuit
JTAG接口电路	JTAG interface circuit
按键电路	key-pressing circuit
数据传输	data transmission
人机交互系统	human-computer interaction system
吸地风扇技术	suction fan technology
占空比	duty cycle
角速度	angular velocity
红外传感器	IR sensor
红外检测电路	infrared detection circuit
红外线	infrared light
红外校准	infrared calibration
红外强度	infrared intensity
红外发射头	infrared transmitter
红外接收头	infrared receiver
红外PWM发生器模块	PWM signal generator driver module
右前方、左前方、左方、前方、右方	the front-right, the front-left，the left, the front and the right
g段	the g segment
空心杯直流电机	coreless DC motor

Appendix C

Course Content and Class Arrangement

Number	Content	Class
Chapter 0　The Origin and Development of Maze Robot	Project 1　The Origin and Development of the Maze Robot Project 2　The Origin of the Maze Robot and the World Vocational College Skills Competition	Theoretical classes: 2
Chapter 1　Basics of Maze Robot Simulation	Project 1　The Basics of Python Project 2　Learning Environment Construction Project 3　Getting Started with ROS2 Project 4　Getting Started with Gazebo	Theoretical classes:2 Practical classes :8
Chapter 2　Maze Robot Simulation Design	Project 1　Learn How to Make a Maze Robot Project 2　Understand the Basic Functions of a Maze Robot Project 3　Master the Intelligent Control of a Maze Robot	Theoretical classes: 6 practical classes: 12
Chapter 3　Maze Robot Simulation Practice	Project 1　Learn the Intelligent Algorithm of a Maze Robot Project 2　The Virtual Simulation Training of the Maze Robot Competition	Theoretical classes: 6 Practical classes: 12
Total		Total classes:48 Theoretical classes:16 Practical classes: 32

Micromouse Technology and Application Series Books
EPIP Teaching Model Series Coursebooks

Simulation and Design of Maze Robot

（A Chinese and English Bilingual Version）

Wang Chao   Hua Yuxiang   Song Lihong◎**Editors in chief**
Gao Yi   Zheng Yongfeng◎**Associate editors**
Liu Jia   Yan Jingyi   Yang Yingdi◎**Translators**

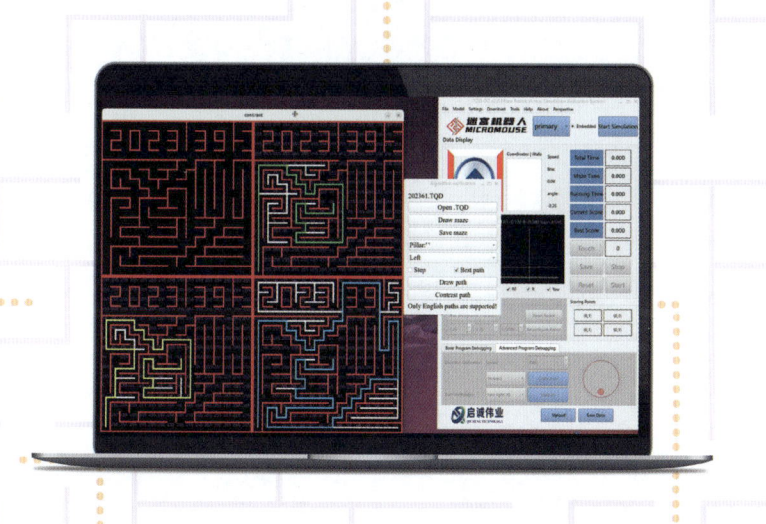

中国铁道出版社有限公司
CHINA RAILWAY PUBLISHING HOUSE CO., LTD.

Introduction

Based on the TQD-OC V2.0 Simulation Evaluation System provided by Tianjin Qicheng Science & Technology Co., Ltd., the book is composed of four chapters, "origin and development", "simulation foundation", "simulation design", and "simulation practice". The book not only expounds on the development, simulation design, development environment, and functional debugging of maze robot but also illustrate the maze robot virtual-reality interaction, intelligent algorithm analysis and application. At the same time, the appendix of this book provides rich teaching resources including the rules of the 2022 World Vocational College Skills Competition—the Maze Robot Competition, international training course standards for "Simulation and 1Design of Maze Robot".

This book is equipped with rich resources such as videos, images, and texts on important knowledge, technical (skill) and literacy key points. Learners can obtain relevant information by scanning the QR code in the book.

This book can be used as a textbook for comprehensive and innovative practical teaching in automation related majors in universities. It can also be used as a textbook for course teaching, vocational enlightenment, scientific and technological activities, and characteristic education in electronic information and automation related majors in vocational colleges. It can also be used as a book for engineering and technical personnel training and a reference book for maze robot enthusiasts.

About the Authors

Wang Chao

Tianjin University, professor, and member of the Teaching Guidance Committee for Automation Majors under the Ministry of Education. He has presided over more than 10 national and provincial level projects, including the national natural science foundation's Special Project of Major Instruments and received 2 first prizes and 3 second prizes of science and technology awards at provincial and ministerial level. The "Virtual Simulation Experiment of Complex System Control for Brewing Process" hosted by Wang was approved as the national first-class course in 2023. He has published 7 textbooks as the first editor in chief, among which the *Micromouse Design Principle and Production Process* (*a Chinese and Bilingual version*) series of textbooks were selected as one of the first batch of national planning textbooks for vocational education during in "14th Five-year Plan" period in 2023. He has also received a second prize of the National Teaching Achievement Award for Higher Education (Undergraduate), a first prize of the Tianjin Teaching Achievement Award, and a second prize of the Higher Education Teaching Achievement Award of the China Association of Automation. Furthermore, he has won other honors, including Baogang Excellent Teacher Award, Ministry of Education New Century Talent Award, and Tianjin Youth Science and Technology Award. He hosted and completed the collaborative education project led by the Ministry of Education, selected as the excellent project case of 2021.

Hua Yuxiang

Tianjin Bohai Vocational Technical College, deputy secretary of the Party Committee and dean. She is responsible for the construction of high-level vocational colleges and professional groups with Chinese characteristics, and has played a leading role in the construction of the Ministry of Education's virtual simulation training base cultivation project. She has hosted the project, "A Research on the Construction Path and Effectiveness of Green Ecological Chemical Virtual Simulation Training Base", approved for the Ministry of Education's Higher Education Science Research and Development Center project. She has presided over the construction of the approved teaching resource database for the chemical safety technology major in Tianjin vocational education, hosted and completed the Tianjin Highly Skilled Talent Training Base and led the construction of six Tianjin training bases approved for the World Skills Competition. She lead the application for the construction of an open petrochemical practice center for the integration of industry and education. Emphasizing the refinement of educational and teaching achievements, she has established the "1+3+4" ideological and political education system, has led the construction of the "dual system" teaching team, and has won multiple teaching achievements at provincial and ministerial level.

Song Lihong

Chairman of Tianjin Qicheng Science & Technology Co., Ltd., founder of Qicheng Micromouse, and senior engineer, with over 20 years dedicated to the software and hardware development, design, production, and service of intelligent micro-motion devices (maze robot). In the past three years, as the leader of the Qicheng technical team, presided over three national and Tianjin Science and Technology Commission research projects. In the last three years, published 3 papers, and edited 6 monographs, including *Micromouse Design Principle and Production Process*(*a Chinese and English bilingual version*) published in 2021, which was selected as one of the first batch of "14th Five-year Plan" national planning textbooks for vocational education. Since 2016, in active response to the national "the Belt and Road" initiative and the "going global" strategy of Chinese enterprises, Qicheng maze robots have been shared with the world through the Luban Workshops abroad. Currently, Luban Workshop schools in Thailand, India, Indonesia, Pakistan, Cambodia, Nigeria, Egypt, and Uzbekistan have adopted *Micromouse Design Principle and Production Process*(*a Chinese and English bilingual version*) as textbooks, offering related international courses to train local talents in intelligent technology along "the Belt and Road" partner country, making efforts and contributions.

Gao Yi

Master's supervisor at the School of Electronic Information Engineering at Nankai University. He is the deputy director of the Electronic Information Experimental Teaching Center at Nankai University. He has participated in multiple projects of "National High tech Research and Development Plan (863 Plan)", key projects of "Tianjin Science and Technology Support Plan", and horizontal scientific research projects. He has published 16 research articles on electronics and educational reform, many of which were indexed by EI. He has also led students to participate in various competitions such as the National College Student Electronic Design Competition, Tianjin College Student IEEE Micromouse Competition, etc. and has achieved excellent results.

Zheng Yongfeng

Dean of the School of Mechanical and Electrical Engineering at Tianjin Bohai Vocational Technical College. He has won the first prize of the Tianjin Teaching Achievement Award, the first prize and the second prize of the China Petroleum and Chemical Education Teaching Achievement Award. The internationalization professional standard for mechatronics integration compiled by him has obtained certification from the Thai Vocational Education Commission. He has published over 10 papers, obtained over 30 patents and has been the chief editor of 5 textbooks.

Liu Jia

PhD from the School of Foreign Languages at Nankai University. She is the leading lecturer of the state-level excellent courses "College English" and "Research Methodology". She has won the first prize in the Nankai University Teaching Skills Competition and the second prize in the Tianjin Teaching Skills Competition. She has compiled oral English textbooks, translation textbooks, writing textbooks, and dictionaries, and has translated several English books. Among them, the translated version of *Micromouse Design Principle and Production Process* (*a Chinese and English bilingual version*) series of textbook were selected as one of the first batch of the national planning textbooks for vocational education in the "14th Five-year Plan" period in 2023.

Yan Jingyi

Master of Communication from Northwestern University in the United States, currently serving as the general assistant of Tianjin Qicheng Science & Technology Co., Ltd. In 2018, she worked as a translator for Cambodian Luban Workshop Micromouse project. During her stay in Cambodia, she tutored local students to learn Chinese Micromouse technology and won the honorary certificate of Cambodian National Institute of Technology. In 2021, she translated the series of *Micromouse Design Principle and Production Process* (*a Chinese and English bilingual version*) textbooks, which was selected as one of the first batch of the national planning textbooks for vocational education during the "14th Five-year Plan".

Yang Yingdi

Teacher at the international exchange and cooperation office of Tianjin Bohai Vocational Technical College. In 2022, she guided students to win two third prizes in the FLTRP Cup English Speaking Contest held by China Education Television, led three bureau-level projects, received funding for one project, and the concluded project results won a second prize in Tianjin.

Maze robots, also known as Micromouse. The technology of maze robots integrates multiple technologies, including intelligent control algorithms, multi-sensor fusion, 3D virtual simulation, intelligent image recognition, internet of things and communications, comprehensively benchmarking the development of modern intelligent technology industry.

This book is one of Micromouse technology and application books, and also a coursebook for the Engineering Practice Innovation Project (EPIP) teaching model. *Micromouse Design Principles and Production Process* (*a Chinese and English bilingual version*) in the book series was selected as the national planning textbook for vocational education in the "14th Five-year Plan" period in 2023, and was introduced to Luban Workshop of "the Belt and Road", making contributions to the training of local intelligent technical talents.

In recent years, the accelerated global digital development, and the widespread application and deep integration of new generation information technology and digital technology are gradually reshaping the concept, mode, and form of education. Digital technologies such as artificial intelligence and virtual simulation are increasingly applied in teaching and competitions, further improving the efficiency, quality, and accuracy of students' knowledge acquisition. Tianjin University, Nankai University, Tianjin Bohai Vocational Technical College and Tianjin Qicheng Science & Technology Co., Ltd. are collaborating to innovate the Micromouse Competition though the Engineering Practice Innovation Project (EPIP) teaching model. They constructed a competition mode that combines online and in-person instruction and virtual simulation and real interaction, stimulating students' awareness of innovation through the competition content and methods, improving teachers' ability to integrate and apply information technology in instruction, empowering vocational education with digital transformation to achieve high-quality

development. It plays a crucial role in the development of the Micromouse Competition and its integration into teaching in the future. In order to further promote and apply the achievements of Micromouse, a book titled *Simulation and Design of Maze Robot* (*A Chinese and English Bilingual Version*)was compiled for higher education and vocational education. The four chapters in the book, "origin and development", "simulation foundation", "simulation design", and "simulation practice", introduce the design, development, functional debugging of the maze robot. The book also expounds on the virtual and real interaction, intelligent algorithm analysis, and application of the maze robot. It is aimed to enrich students' engineering practice innovation ability, expand their professional horizons, and contribute to mastering professional literacy, and assist in the digital transformation and development of teaching.

The book is characterized by the following features:

1. The two-way empowerment of school and enterprise fully reflects the cooperative concept of industry education integration

The textbook is designed in line with students' professional ability growth. The teaching staff from college and university and the enterprise technical personnel jointly formulate curriculum standards and compile the book. The selected cases in this book are all from real engineering projects, and the editors are from universities and enterprises in China and they have long been engaged in maze robot design and development, and have led students to win awards in international Micromouse competitions. Especially in August 2022, at the first World Vocational College Skills Competition, Tianjin Bohai Vocational Technical College hosted the Micromouse Competition and Tianjin Qicheng Science & Technology Co., Ltd. was selected as a cooperative enterprise. They cooperated and jointly designed the TQD virtual simulation Micromouse Competition evaluation system, which has been widely applied in the competition and has received recognition from experts. This system follows international competition standards, and innovation and reforms are made based on the actual situation in education and teaching, formulating virtual

simulation Micromouse Competition standards with Chinese characteristics.

2. Seize the new opportunity of Luban Workshops of partner countries along "the Belt and Road" to promote the internationalization of vocational education

"Promoting the joint construction of high-quality development along 'the Belt and Road'" is an important measure proposed in the report of the 20th CPC National Congress. Building an international brand of Chinese vocational education is an important task in promoting the reform of the modern vocational education system. By 2023, the bilingual maze robot textbooks have been used in the course construction for nine "Luban Workshops" in eight countries, including Thailand, India, Indonesia, Pakistan, Cambodia, Nigeria, Egypt and Uzbekistan. The international curriculum standards in the book are co-established by Tianjin Bohai Vocational Technical College and Tianjin Qicheng Science & Technology Co., Ltd. The maze robot software and hardware teaching equipment is spread in the world along with Luban Workshop. The integration of industry and education is significant, providing rich practical teaching resources for partner countries along "the Belt and Road", and being of help for the training of local technical talents.

The chief editors are professor Wang Chao from Tianjin University, Hua Yuxiang, deputy secretary of the Party Committee and dean of Tianjin Bohai Vocational Technical college, Song Lihong, general manager of Tianjin Qicheng Science & Technology Co., Ltd. and founder of Qicheng Micromouse. The associate editors are Gao Yi, associate professor from Nankai University, professor Zheng Yongfeng from Tianjin Bohai Vocational Technical College. The translators are Liu Jia, from the School of Foreign Languages at Nankai University, Yan Jingyi, general assistant at Tianjin Qicheng Science & Technology Co., Ltd., and Yang Yingdi, from Tianjin Bohai Vocational Technical College.

When compiling the book, we consulted many references and received support from professors and experts from universities such as Tianjin

University, Tianjin Bohai Vocational Technical College, and Nankai University. At the same time, Tianjin Qicheng Science & Technology Co., Ltd. provided practical engineering cases, mind maps, QR code videos, and animation course resources. Furthermore, the book is published by China Railway Publishing House Co., Ltd. We would like to express our sincere gratitude to them.

Although we have made some breakthroughs in the textbook construction, there must be some mistakes and omissions in the book due to the editors' limited knowledge and time. We sincerely ask for instruction from all experts and readers.

Wang Chao　Hua Yuxiang　Song LiHong

January, 2024

Chapter 0   The Origin and Development of Maze Robot 001

Project 1   The Origin of the Development of Maze Robot 003

Project 2   The Origin of the Maze Robot and the World Vocational College Skills Competition 011

Chapter 1   Maze Robot Simulation Foundation 015

Project 1   The Basics of Python 017

   Task 1   Get to Know the Python Development Environment 017

   Task 2   Master Python Data Types 029

   Task 3   Learn Python Functions and Classes 043

   Task 4   Master Python Extension Libraries................................ 049

Project 2   Learning Environment Construction............................... 054

   Task 1   Getting Started with the Ubuntu System............................ 054

   Task 2   Learn ROS2 Installation.. 063

   Task 3   Learn Gazebo Installation and Integration....................... 067

Project 3   Getting Started with ROS2 072

   Task 1   Understand the ROS2 Workspace and Common Commands 072

   Task 2   Learn ROS2 Nodes 077

   Task 3   Learn ROS2 Topic.. 080

Project 4   Getting Started with Gazebo 089

   Task 1   Master the Basic Operations of Gazebo............................ 089

   Task 2   Learn to Build URDF Model 095

Chapter 2   Maze Robot Simulation Design 103

Project 1   Learn How to Make a Maze Robot............................... 105

   Task 1   Make a Virtual Simulation Maze Robot............................ 105

Task 2　Create a Virtual Simulation Maze 110

Task 3　Learning Virtual Simulation System Startup 115

Project 2　Understand the Basic Functions of a Maze Robot 119

Task 1　Understand Two-wheel Differential Drive 119

Task 2　Learn Lidar Detection 129

Task 3　Master the Maze Robot's Tracking Operation 134

Project 3　Master the Intelligent Control of a Maze Robot 138

Task 1　Learn Coordinate Calculations.. 138

Task 2　Learn Intersection Detection 143

Task 3　Master Turning Control.. 145

Chapter 3　Maze Robot Simulation Practice 153

Project 1　Learn the Intelligent Algorithm of a Maze Robot...................... 155

Task 1　Understand the Maze Search Strategy 155

Task 2　Master the Storage and Reading of Maze Information.............. 160

Task 3　Master the Optimal Path Planning of the Maze 164

Project 2　The Virtual Simulation Training of the Maze Robot

Competition 168

Task 1　Learn the Virtual Simulation Evaluation System........................ 168

Task 2　Understand the Virtual Simulation Competition........................ 175

Appendix 179

Appendix A　The Rules of the 2022 World Vocational College Skills

Competition — the Maze Robot Competition 179

Appendix B　International Training Course Standards for "Simulation and

Design of Maze Robot" 185

Appendix C　Course Content and Class Arrangement.................................. 188

Appendix D　Chinese-English Comparison Table of Professional

Vocabulary.. 189

Chapter 0

The Origin and Development of Maze Robot

Maze robots, also known as Micromouse. The technology of maze robots integrates many related technologies, including intelligent control algorithm, multi-sensor fusion, 3D virtual simulation, intelligent image recognition, internet of things and communications, fully reflecting the development of modern intelligent technology industry. The maze robot competition is an important medium in which the results of industry-university-research cooperation can be demonstrated and the transformation of competition achievements can be promoted.

The maze robot competition has a history of more than 50 years in the world, including classical reality competition and virtual simulation competition. The maze robot competition establishes a hybrid competition mode that combines online and offline as well as virtual simulation and reality interaction. It innovates the content and methods of the maze robot competition, stimulates the vitality of college students, and empowers the high-quality development of higher education and vocational education through digital transformation.

The main content of the classical reality competition is that a maze robot starts from the starting point, searches the maze independently in the unknown maze without human manipulation, finds the end point, and selects the shortest path to sprint. The competitors are ranked according to the searching time and the sprint time. The maze setting follows the international standards of the Institute of Electrical and Electronics Engineers (IEEE).

With the acceleration of the global digital process, the new generation of information technology and digital technology are widely used and deeply integrated, gradually "reshaping" the concept, mode and form of education.

Under the circumstances, Tianjin Qicheng Science & Technology Co., Ltd. cooperates with Tianjin Bohai Vocational Technical College. Following the international competition standards and making innovations according to the actual teaching situation, they jointly develop and design the TQD virtual simulation maze robot competition evaluation system, and establish the virtual simulation maze robot competition standards with Chinese characteristics.

The virtual simulation contest requires contestants to first build a 3D maze and 3D robot environment in the computer system, deploy the virtual simulation contest environment, and then a 3D robot traverses the complex maze, calculates and evaluates the optimal path through the intelligent control algorithm, and sprints from the starting point to the finishing point at the fastest speed.

The maze robot competition is oriented to the development needs of many industries, such as electronic information technology and equipment manufacturing. Focusing on intelligent control application and considering international standards, professional standards and job requirements of information technology and intelligent control technology, the competition aims at assessing the knowledge and skills involved in the implementation of intelligent control projects. The competition integrates the key technologies, core knowledge and core skills of embedded microcontroller technology and application, sensor technology and application, intelligent control algorithm application, 3D virtual simulation technology application, artificial intelligence and computing technology application. The event fully demonstrates new technological achievements in artificial intelligence and robotics, orients the training direction of advanced technical and skilled talents, leads professional upgrading of vocational colleges, improves the ability of professional construction, promotes the transformation of competition achievements and the international cooperation among industry, education and research institutes, closely follows the market demand, focuses on new technologies and new formats, builds a high-level platform for international vocational and technical education exchanges, and supports the high-quality development of vocational and technical colleges, transmits the educational philosophy of China's vocational education, exports the experience of China's vocational education competition, and strengthens the international communication and cooperation.

The book systematically introduces the maze robot technology from 3D modeling, software development environment and programming, and specifically explains the basic principles and practical operation methods of the maze robot.

Chapter0   The Origin and Development of Maze Robot   003

Project 1

The Origin of the Development of Maze Robot

Learning objectives

1. Understand the origin of the maze robot.

2. Understand the development of domestic and international maze robot competitions.

In 1938, Shannon, a mathematician from Michigan, USA, completed the paper *Symbolic Analysis of Relay and Switch Circuits*. Since Boolean algebra has only 0 and 1, which just corresponds to binary, Shannon applied it to relay switches that process information in pulse mode, completely changing the design of digital circuits both theoretically and technologically. Therefore, this paper has epoch-making significance in the history of modern digital computers.

In 1948, Shannon published another paper, *Mathematical Theory of Communication*, and won himself the title of "Father of Information Theory".

In 1956, he participated in launching the Dartmouth Artificial Intelligence Conference, becoming one of the founders of this new discipline. He not only took the lead in applying artificial intelligence to computer chess, but also invented a maze robot that can automatically pass through the maze so as to prove that computers can improve their intelligence through learning.

The origin and development of maze robot is shown in Figure 0-1-1.

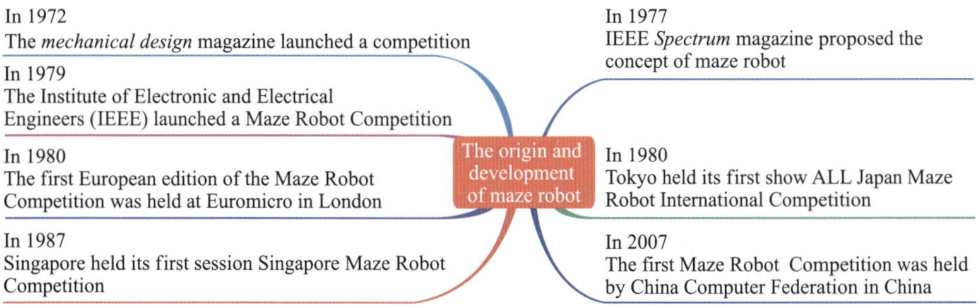

In 1972
The *mechanical design* magazine launched a competition

In 1977
IEEE *Spectrum* magazine proposed the concept of maze robot

In 1979
The Institute of Electronic and Electrical Engineers (IEEE) launched a Maze Robot Competition

In 1980
The first European edition of the Maze Robot Competition was held at Euromicro in London

The origin and development of maze robot

In 1980
Tokyo held its first show ALL Japan Maze Robot International Competition

In 1987
Singapore held its first session Singapore Maze Robot Competition

In 2007
The first Maze Robot Competition was held by China Computer Federation in China

Figure 0-1-1   The origin and development of maze robot

In 1972, the *mechanical design* magazine launched a competition. In the competition, mechanical mouses driven only by the spring of the mousetrap compete to cover the longest distance along the runway.

In 1977, IEEE *Spectrum* magazine proposed the concept of maze robot, a small robot vehicle controlled by microprocessor, which can decode and navigate in complex mazes.

In 1979, the Institute of Electrical and Electronics Engineers (IEEE) initiated a maze robot competition through the IEEE *Spectrum* magazine, offering a $1000 prize to the designer of a maze robot that could autonomously navigate out of a maze in the shortest time.

In 1980, the first World Maze Robot Competition was held in Japan. After that, many other competitions were held, such as the UK Maze Robot Competition in 1980, the IES Maze robot Competition in Singapore in 1987, and the Maze Robot Competition held by China Computer Federation in China in 2007.

From the mechanical and electronic mouse in 1972 to the present one, as shown in Figure 0-1-1, the maze robot competition has undergone more than 50 years of transformation. The contestants were originally graduate students from Harvard University, Massachusetts Institute of Technology and other world famous universities at the beginning, but now they can be students from universities, applied technology universities, vocational colleges and even primary and secondary students. maze robot is adopted as the teaching carrier at multiple educational levels to cultivate students' engineering literacy, scientific and technological innovation awareness and practical design ability. Various maze robot competitions have also sprung up and flourished. Up till now, maze robot competitions have been applicable to students at different educational stages in the world.

Since 2007, the maze robot has experienced more than 10 years of development in China, as shown in Figure 0-1-2. In 2007, Tianjin Qicheng Science & Technology Co., Ltd. introduced this international event to Tianjin, China. With the core concept of China's advanced education model "Engineering Practice Innovation Project", the company carried out a localized innovation reform, supported the maze robot competition to flourish in China, and played a key role in introducing the maze robot technology in classroom teaching.

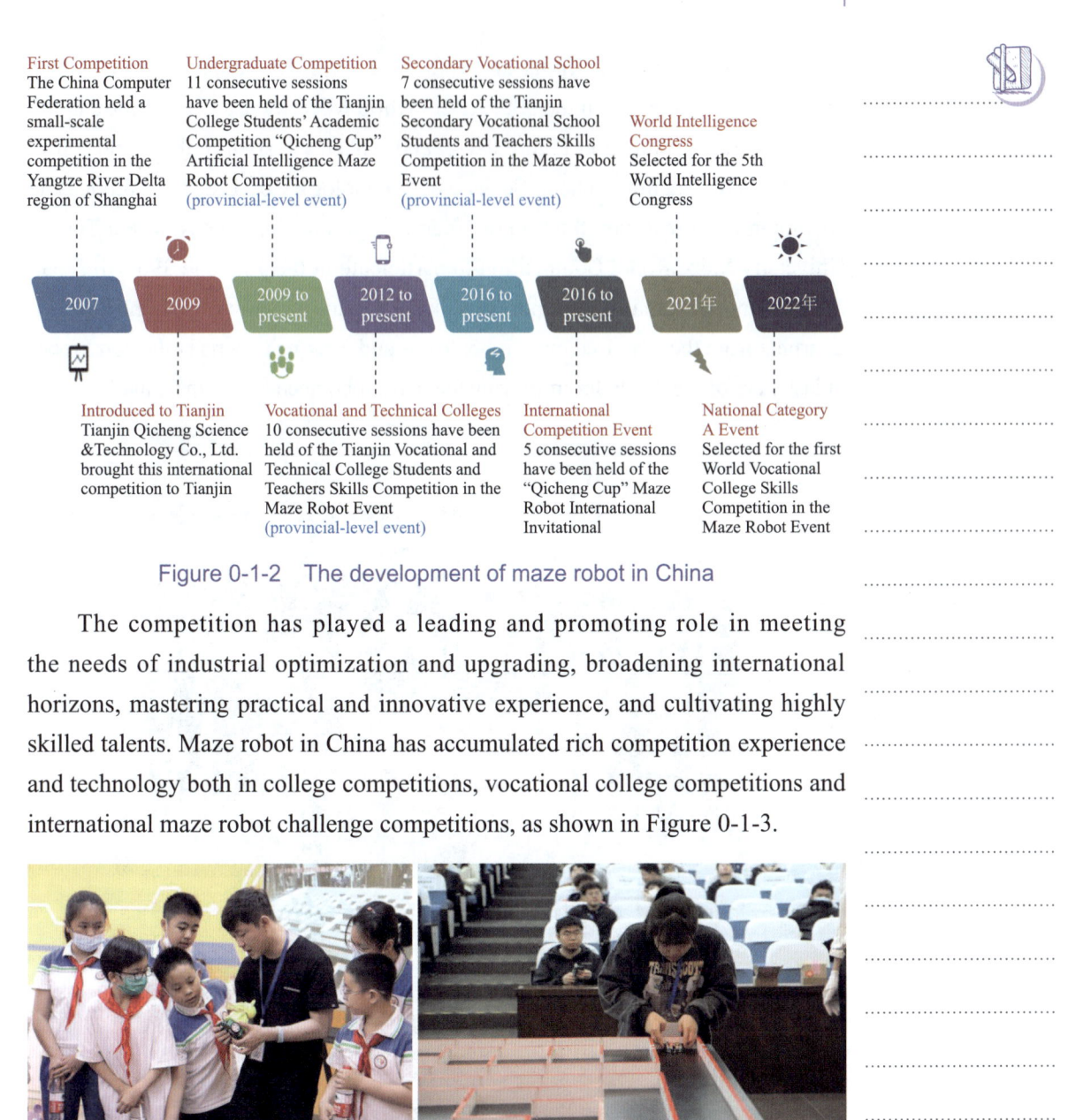

Figure 0-1-2　The development of maze robot in China

　　The competition has played a leading and promoting role in meeting the needs of industrial optimization and upgrading, broadening international horizons, mastering practical and innovative experience, and cultivating highly skilled talents. Maze robot in China has accumulated rich competition experience and technology both in college competitions, vocational college competitions and international maze robot challenge competitions, as shown in Figure 0-1-3.

Figure 0-1-3　Competition photos

　　For more than a decade, China's maze robot competition has constantly innovated for international development. From the initial "simple imitation" to current "mutual learning", a new platform for international exchange and cooperation has been built. There are three stages of development: learning and reference, transformation and sublimation, and guidance and radiation.

The first stage is learning and reference. In 2015, students from Tianjin universities participated in the 30th APEC World Micromouse Competition in the United States, as shown in Figure 0-1-4, and ranked the sixth in the world. From 2017 to 2018, Tianjin Qicheng Science & Technology Co., Ltd. fully funded the champion team that won the title of enterprise proposition contest in the Tianjin University Maze Robot Competition to participate in the 38th and 39th All Japan Maze Robot International Open in Tokyo, Japan, as shown in Figure 0-1-5. Learning from the advanced maze robot teams and meeting experts in the maze robot industry are of great help for improving the maze robot technology in China.

Figure 0-1-4　Tianjin teams participating the contest in America

Figure 0-1-5　Tianjin teams participating the contest in Japan

Chapter0　The Origin and Development of Maze Robot

The second stage is transformation and sublimation. At the stage, the maze robot competition has experienced innovation in China. A series of Qicheng maze robot teaching platforms, ranging from easy level to difficult level, have been designed to meet the needs of students at different learning stages. As shown in Figure 0-1-6 to Figure 0-1-8, since 2016, maze robot experts, including professor David Otten of Massachusetts Institute of Technology, professor Su Jinghui of Longhua University of Science and Technology in Taiwan, China, professor Huang Mingji of Yi'an Institute of Technology in Singapore, Professor Peter Harrison of Birmingham City University in the United Kingdom, Mr Yoriko Nakagawa, Secretary General of the Organizing Committee of the Japan Maze Robot International Championship, teachers and students of the international Luban Workshop from Thailand, India, Indonesia, Pakistan and Cambodia, as well as the elite teams from Tianjin, Beijing, Henan, Hebei and other provinces and cities, have successively joined in the Maze robot International Invitational Tournament in China. By participating in China's competitions, international players have a deeper understanding and recognition of China's competition standards, rules, modes and concepts, so as to effectively promote international exchanges and cooperation and achieve the goal of "mutual learning and mutual learning".

Figure 0-1-6　The third IEEE Micromouse International Invitation Contest

Figure 0-1-7　The fourth "Qicheng Cup" IEEE Micromouse International Invitation Competition

Figure 0-1-8　The fifth "Qicheng Cup" Micromouse International Invitation Competition

　　The third stage is guidance and radiation. Education opening up is an important part of China's reform and opening up cause. Since 2016, with the promotion of "the Belt and Road" initiative, under the guidance of the Ministry of Education, the international project of Luban Workshop has been launched. As China's excellent educational equipment, the maze robot has been shared with the world along with the promotion of Luban Workshop. As shown in Figure 0-1-9 to Figure 0-1-14, since 2016, Qicheng maze robot has been introduced to Thailand, India, Indonesia, Pakistan, Cambodia, Nigeria, Egypt, Uzbekistan and other overseas countries, the competition has been promoted and relevant course

Chapter0　The Origin and Development of Maze Robot　009

training has been set up, all of which have been unanimously favored by teachers and students in these countries. In this way, the maze robot has become the link connecting the world.

Figure 0-1-9　Maze robot training course at Luban Workshop in India

Figure 0-1-10　Maze robot training course at Luban Workshop in Thailand in 2016

Figure 0-1-11　Maze robot training course at Luban Workshop in Indonesia in 2017

Figure 0-1-12　Maze robot training course at Luban Workshop in Pakistan in 2018

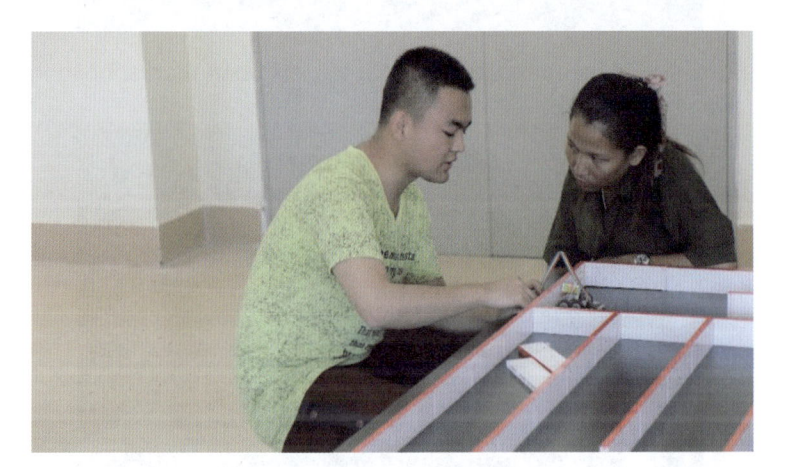

Figure 0-1-12　Maze robot training course at Luban Workshop in Cambodia in 2018

Figure 0-1-14　Maze robot training course at Luban Workshop in Egypt in 2020

Project 2

The Origin of the Maze Robot and the World Vocational College Skills Competition

Learning objectives

Understand the origin of the maze robot and the World Vocational College Skills Competition.

In 2022, the Ministry of Education held the World Conference on the Development of Vocational and Technical Education in Tianjin, the first international conference on vocational education hosted by the Chinese government. At the same time, the World Vocational College Skills Competition was held. The latter is held every two years on the principles of "enhancing exchanges, deepening cooperation and innovative-driven development". The conference has important international influence in the field of vocational education. It aims at promoting the mutual exchange among global vocational education and building a community of shared future for mankind.

In August 2022, the executive committee of the World Vocational College Skills Competition announced *the Notice on Announcing the Organizers and Cooperative Enterprises of the First World Vocational College Skills Competition* (SEWH [2022] No. 38). In the document, Tianjin Bohai Vocational Technical College was selected as the organizer of the maze robot competition event. Tianjin Qicheng Science & Technology Co., Ltd. was selected as a cooperative enterprise in the maze robot competition event, as shown in Figure 0-2-1.

By 2022, relevant professional courses on the maze robot have been taught and EPIP training centers established in nine Luban Workshops in eight countries, including electromechanical integration technology major in Thailand, industrial robot major in India, electronic technology application major in Indonesia, automation control major in Pakistan, mechatronics technology major in Cambodia, electrical and electronic engineering major in Nigeria, industrial

robot major and automatic control major in Egypt and professional courses on information technology in Uzbekistan. The maze robot has the characteristics of multi-specialty group construction and application and multi-industry scene focused application. It can assess the knowledge and skills involved in the implementation of intelligent control projects according to the international standards, professional standards and job requirements of information technology and intelligent control technology.

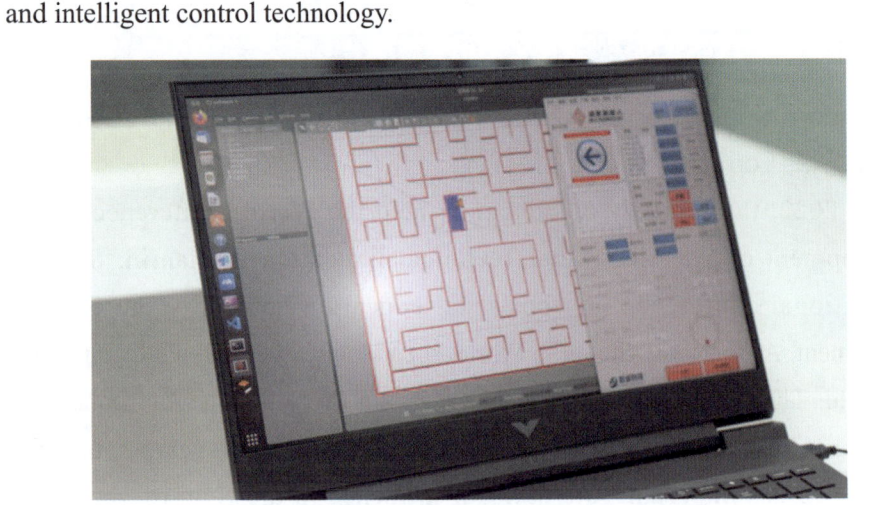

Figure 0-2-1   "Maze robot" event of the first World Vocational College Skills Competition

On August 11, 2022, the maze robot competition, one event of the first World Vocational College Skills Competitions, kicked off. The skills competitions were hosted by 35 units, including Ministry of Education, National Development and Reform Commission, Ministry of Science and Technology, Ministry of Industry and Information Technology, and Tianjin People's Government, organized by Tianjin Municipal Education Commission, Tianjin Academy of Education Sciences and Tianjin Bohai Chemical Group Co., Ltd., and co-organized by Tianjin Bohai Vocational Technical College and Tianjin Qicheng Science & Technology Co., Ltd. Players from both China and other countries, including Equatorial Guinea, Benin, Yemen, Ethiopia, Sudan, Congo, Sierra Leone, Zimbabwe, competed fiercely online.

Finally, the team composed of Kapushi Saignon John, a contestant from the Republic of Benin in West Africa (see Figure 0-2-2), and Yu Xinling, a student

Chapter0   The Origin and Development of Maze Robot   013

from Tianjin Bohai Vocational Technical College, won the gold medal.

In an interview, John revealed that he was curious about science and technology since he was young. During his study in Tianjin, he felt very lucky to have good teachers. Cooperating with Yu Xinling, he presented a miniature version of "speed and passion" for the spectators. He held that the contest did not only improve their skills, but also enhance the mutual exchange. He firmly believed that China's vocational and technical education would increasingly develop and bring more value to the world.

Figure 0-2-2   One contestant of the maze robot event

Gao Yi, associate professor of Nankai University and the event referee, said that in recent years, 3D virtual simulation technology had attracted much attention. The virtual simulation competition in 2022 was favored by international teachers and students. The network makes it possible for players to join in a fair, just and open competition in real time through the TQD-OC V2.0 virtual simulation evaluation system, regardless of geographical and temporal constraints. Maze robots are gradually upgrading from intelligent motion devices to digital twin technology. Guided by the service industry and technology development, they actively integrate with international vocational education. With Luban Workshop as a link, maze robots serve "the Belt and Road" construction, deepen industry-university-research institute cooperation, aim at market demand, and further build a national vocational education highland.

Questions and Summary

1. Which two events are included in the maze robot competition? What are the differences between them?

2. What is the function of the virtual simulation evaluation system for maze robot?

3. The maze robot competition has evolved into an international innovative student competition applied at different educational stages. As a teaching medium, the maze robot fosters engineering literacy, technology innovation awareness, and hands-on design skills among students.

Chapter 1

Maze Robot Simulation Foundation

Maze robot simulation is the use of software and programs to simulate the entire operation process of the maze robot. In this Chapter, it mainly explains the programming language and software running environment used in simulation.

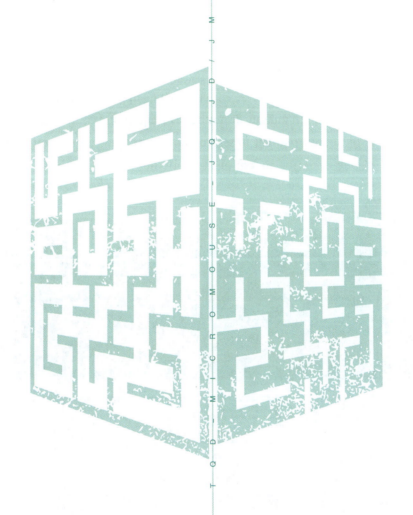

Chapter1 Maze Robot Simulation Foundation | 017

Project 1

The Basics of Python

Learning objectives

1. Understand how to install and use Python and VSCode.

2. Understand the basic syntax of Python.

3. Understand commonly used Python standard libraries and third-party libraries.

Task 1   Get to Know the Python Development Environment

This task involves learning how to set up the Python development environment and mastering basic syntax.

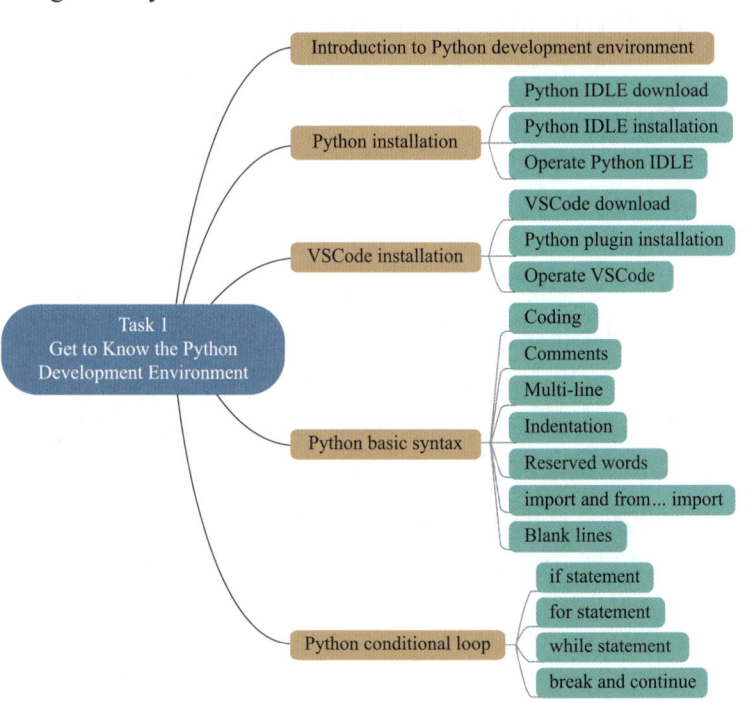

I. Introduction to Python development environment

Python is a cross-platform computer programming language. It is an object-oriented, interpreted, and dynamically typed high-level language that can be used on Windows, Linux, and Mac platforms, as shown in Figure 1-1-1.

Figure 1-1-1　Python icon

Python has simple syntax and is easy to use. Even beginners who are not software majors can easily get started. More and more people are starting to use Python for software development. Compared with other programming languages, Python has the following advantages:

① Free and open-source. Developing or releasing programs with Python does not require any payment, and there are no concerns about copyright issues. Even for commercial purposes, Python is still free.

② Syntax is simple. Compared with C/C++, C#, VB, Java and other languages, Python has lower requirements on code format, and you don't need to spend too much effort on details when writing programs.

③ High level programing language. It is an encapsulation of the C language and shielding many low-level details. Such as automatic memory management. It is automatically allocated when needed and freed when not needed.

④ Good portability and can be applied to multiple platforms.

⑤ An object-oriented programming language that features object-oriented characteristics such as inheritance, polymorphism, and encapsulation.

⑥ Strong extensibility and a wide range of extension libraries that can assist in completing various types of programs. It covers a vast majority of application scenarios, including file I/O, numerical computing, GUI programming, network programming, and database access.

The development history of Python has gone through Python 2 and Python 3. Due to significant differences between them, the official stopped the maintenance of Python 2 in 2020. Therefore, we recommend using Python 3 for development. The Python mentioned in the subsequent content of this book refers to Python 3.

II. Python installation

Python is open source and free. Users can go to the Python official website to download and install it directly.

Chapter1　Maze Robot Simulation Foundation　019

1. Python IDLE download

At present, the latest official version of Python is 3.11.x. When learning Python, it is recommended to use the latest official version. However, when developing a project, it is recommended to lower the version and choose 3.10.x. This is because some third-party libraries may not be adapted yet. The latest official version is prone to various errors.

Choose the appropriate installer based on your computer's operating system, as shown in Figure 1-1-2.

Files

Version	Operating System	Description	MD5 Sum	File Size	GPG	Sigstore
Gzipped source tarball	Source release		7e25e2f158b1259e271a45a249cb24bb	26085141	SIG	.sigstore
XZ compressed source tarball	Source release		1bf8481a683e0881e14d52e0f23633a6	19640792	SIG	.sigstore
macOS 64-bit universal2 installer	macOS	for macOS 10.9 and later	f5f791f8e8bfb829f23860ab08712005	41017419	SIG	.sigstore
Windows embeddable package (32-bit)	Windows		fee70dae06c25c60cbe825d6a1bfda57	7650388	SIG	.sigstore
Windows embeddable package (64-bit)	Windows		f1c0538b060e03cbb697ab3581cb73bc	8629277	SIG	.sigstore
Windows help file	Windows		52ff1d6ab5f300679889d3a93a8d50bb	9403229	SIG	.sigstore
Windows installer (32 -bit)	Windows		83a67e1c4f6f1472bf75dd9681491bf1	27865760	SIG	.sigstore
Windows installer (64-bit)	Windows	Recommended	a55e9c1e6421c84a4bd8b4be41492f51	29037240	SIG	.sigstore

Figure 1-1-2　Python download

Please search for the download address in the search engine by yourself.

It should be noted that in Ubuntu 22.04, 3.10.6 has been integrated and no additional installation is required.

2. Python IDLE installation

The installation file of Python is very small, only about 27 MB. As a demonstration, here select "Windows installer (64-bit) " to install in Windows.

The installation process of Python is similar to that of ordinary software, as shown in Figure 1-1-3. It is recommended to check the two selection boxes.

It is recommended to disable the path length limit, as shown in Figure 1-1-4. Otherwise, failure may occur due to the excessive length of some program file paths.

3. Operate Python IDLE

Python will not create a desktop shortcut. After installation, you can find Python in the menu, as shown in Figure 1-1-5. Click IDLE to start Python, as shown in Figure 1-1-6.

Type the program in IDLE and press【 Enter 】button to execute.

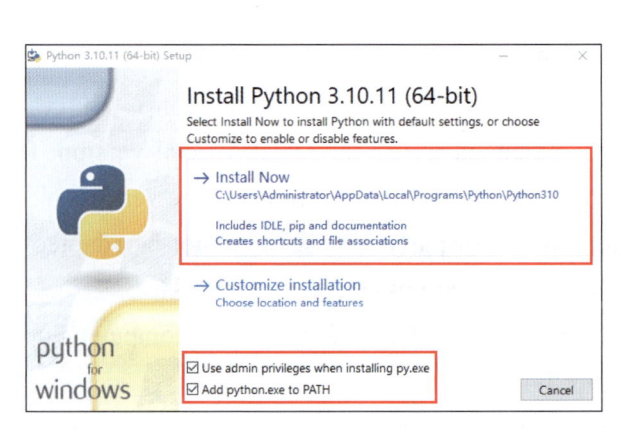

Figure 1-1-3　Check administrator privileges and add path

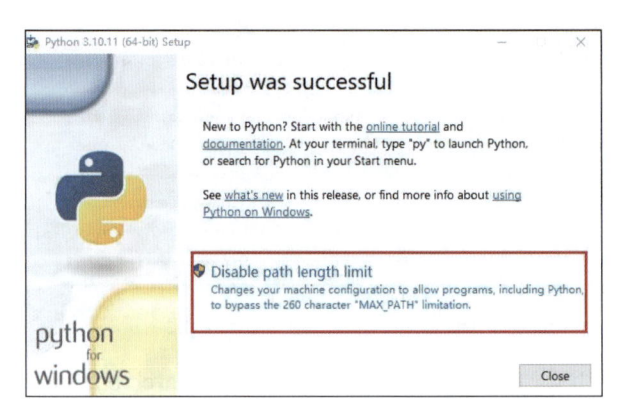

Figure 1-1-4　Disable path length limit

Figure 1-1-5　Find Python

Figure 1-1-6　Use Python IDLE

Ⅲ. VSCode installation

If you are just writing a simple program, you can write the code in the development environment that comes with Python. But for professional

Chapter1　Maze Robot Simulation Foundation | 021

programmers, the programs they write are relatively complex, and it is a bit difficult to write code in the development environment that comes with Python, especially when writing object-oriented programs, whether it is code prompt function or error message. The prompt function is far less powerful than that of a professional development environment.

VSCode is a lightweight editor produced by Microsoft. It is just a text editor. All functions exist in the form of plug-in extensions. If you need any functions, install the corresponding extensions, including Python, HTML, C/C++, database, etc., very convenient. VSCode supports all major operating systems, including Windows, Linux and MacOS. Therefore, this book chooses VSCode as the main editor.

1. VSCode download

VSCode is completely free. You can visit the official website and then download and install it based on your computer's operating system, as shown in Figure 1-1-7.

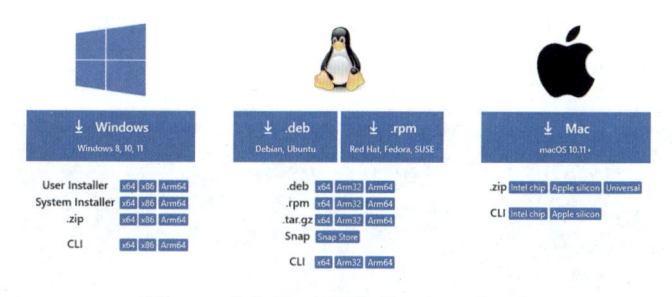

Figure 1-1-7　VSCode download

For Windows users:

If you are a regular user, please select x64, x86 or Arm64 in "User Installer";

If you are an administrator user, please select x64, x86 or Arm64 in the "System Installer".

2. Python plugin installation

For Ubuntu users: You can also go to the system's software store and search for "Code" to download and install.

The installation process of VSCode is the same as that of ordinary software, which is relatively simple and will not be explained here.

To develop Python, you need to install Python extensions, as shown in Figure 1-1-8, by searching for "python" in the plugins.

Figure 1-1-8　Install Python extensions

After installation, you will have the capability for Python development, such as debugging, syntax highlighting, code navigation, intelligent suggestions, auto-completion, unit testing, version control, etc., which greatly improves Python development efficiency.

There are many plug-ins to assist Python development, such as indent-rainbow, Jupyter, Image preview, etc.

3. Operate VSCode

VSCode will automatically identify the Python interpreter on your computer. Create a script file demo.py, type program, and choose to run the Python file in the terminal, as shown in Figure 1-1-9.

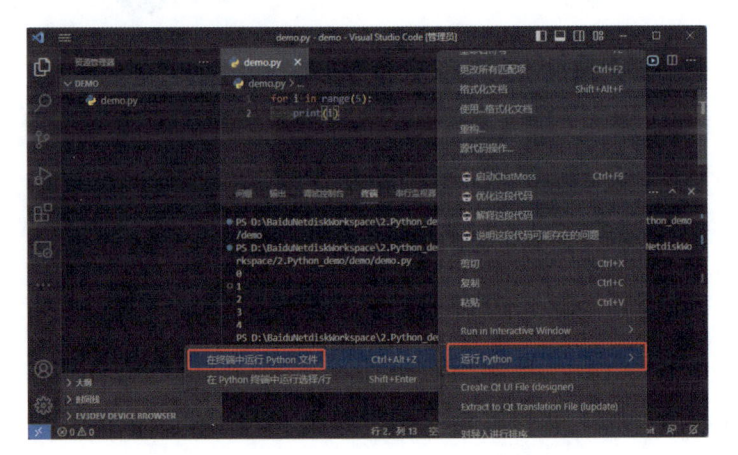

Figure 1-1-9　Run VSCode

After Python and VSCode are installed, you can start basic programming in Python.

Ⅳ. Python basic syntax

Like other programming languages, writing Python programs needs to follow Python syntax rules. The Python language has many similarities with

Chapter1   Maze Robot Simulation Foundation   023

languages such as C and Java. However, there are some differences. The following introduces some basic commonly used syntax.

1. Coding

By default, Python source files are encoded in UTF-8, and all strings are unicode strings. Of course, you can also specify different encodings for the source code files.

```
# -*- coding: cp-1252 -*-
```

2. Comments

Comments are explanations of the code, and their purpose is to make it easier for people to understand the code.

Single-line comments in Python start with #, and multi-line comments are enclosed by three pairs of single quotes (' ' ') or three pairs of double quotes ("""). The Python interpreter will automatically skip this part and will not execute it.

```
if 1+3>2:  # This is a single-line comment
    """
    This is a multi-line comment
    This is a multi-line comment
    """
    '''
    You can also use three single quotes
    to make multi-line comments
    '''
    print(1)
else:
print(0)
```

3. Multi-line

Python is an interpreted language, and programs are generally executed line by line, with a new line serving as the end of the previous line's statement. If the code is too long or for aesthetic reasons, you can split a single line into multiple lines using a backslash "\" to indicate that it is a single statement. Here, the backslash "\" is called a continuation character

```
>>> a = 8+9+\
        10+12+\
        15
>>>a
    54
```

If multiple lines are contained in "[]", "() ", "{}", then there is no need to use connectors.

```
>>> a = (8+9+
        10+12+
        15)
>>>a
    54
```

4. Indentation

The most distinctive feature of Python is the use of indentation to represent code blocks. Other programming languages often use curly brackets "{}". The number of indented spaces is variable, but statements in the same code block must contain the same number of indented spaces.

```
if True:
    print(True)
else:
    print(False)
```

Python officially recommends using 4 spaces for each level of indentation.

```
while 1:
    if 1:                #1 level indent, 4 spaces
        if 1:            #2 levels of indentation, 8 spaces
            print(1)     #3 levels of indentation, 12 spaces
        else:
            print(0)
    else:
        print(0)
```

5. Reserved words

Reserved words are keywords and we cannot use them as any identifier names. Python's standard library provides a keyword module that we can use to view all reserved words in the current version.

```
>>> import keyword
>>> keyword. kwlist
```

The following information is output in the terminal.

```
['False', 'None', 'True', 'and', 'as', 'assert', 'async', 'await',
'break', 'class', 'continue', 'def', 'del', 'elif', 'else', 'except',
'finally', 'for', 'from', 'global', 'if', 'import', 'in', 'is', 'lambda',
```

Chapter1　Maze Robot Simulation Foundation | 025

```
'nonlocal', 'not', 'or', 'pass', 'raise', 'return', 'try',
'while', 'with', 'yield']
```

Python is case-sensitive. For example, FALSE is not a reserved word.

```
>>> False
    False
>>> FALSE
Traceback (most recent call last)    :
  File "<pyshell#33>", line 1, in <module>
    FALSE
NameError: name 'FALSE' is not defined. Did you mean: 'False'?
>>> FALSE=1
>>> FALSE
    1
```

6. import and from...import

Use import or from...import in Python to import extension libraries.

Import the entire module (somemodule) in the format: import somemodule.

Import a function from a module in the format: from somemodule import somefunction.

Import multiple functions from a module in the format: from somemodule import firstfunc, secondfunc, thirdfunc.

Import all functions in a module in the format: from somemodule import *.

You can also use as to represent an alias, in the format: from somemodule import somefunction as newname.

```
>>> import os
>>> os. getcwd()
    'D:\\Python3'
>>> from sys import getdefaultencoding
>>> getdefaultencoding()
    'utf-8'
>>> import numpy as np
>>> np. ones((3,3))
    array([[1., 1., 1.],
           [1., 1., 1.],
           [1., 1., 1.]])
```

7. Blank lines

Blank lines are not part of Python syntax. Even if no blank lines are inserted when writing, the Python interpreter will run without errors. However, the role of a blank line is to separate two pieces of code with different functions or

meanings, making it easier to maintain or reconstruct the code in the future.

Using a blank line indicates the start of a new section of code. It's recommended to use two blank lines to separate classes and functions that are not related, and one blank line to separate methods within a class, for better readability.

```python
def func1():
    pass


def func2():
    pass


class Demo:

    def func1():
        pass

    def func2():
        pass
```

V. Python conditional loop

Python conditional statements are blocks of code that are executed based on the execution results (True or False) of one or more statements.

Python loop statements determine whether to loop through the code block based on the execution result of the conditional statement (True or False).

1. if statement

The if statement is a conditional statement used to form conditional control. It is used as follows:

```python
if condition1:
    statement_block_1
elif condition2:
    statement_block_2
else:
    statement_block_3
```

The if statement will always execute only one statement_block, in which both the elif and else parts can be omitted, or there can be multiple elif, but there can only be one else at most.

The execution order of an if statement is top-down, evaluating the

conditions. As soon as a condition is found to be True, its corresponding statement_block is executed, and other conditions will not be evaluated.

```
a=8
if a>10:
    print('a>10')
elif 5<=a<=10:
    print('5<=a<=10')
else:
    print('a<5')
```

The following information is output in the terminal.

```
5<=a<=10
```

The commonly used operators in an "if" statement, as shown in Table 1-1-1.

Table 1-1-1　The commonly used operator in an if statement

Operator	Description	Operator	Description
<	less than	>=	greater than or equal to
<=	less than or equal to	==	equal to, compares two values for equality
>	greater than	!=	not equal to

2. for statement

The for statement is a loop statement. The loop can traverse any iterable object, such as a list or a string. It is usually used in conjunction with "in". It is used as follows:

```
for variable in iterable:
    statements
else:
    statements
```

The for statement can be used alone or in conjunction with an "else" clause. Unlike the if statement, the "else" block will only execute after the "for" loop has successfully iterated through all items.

```
sites = ["Baidu", "Google","Jingdong","Taobao"]
for site in sites:
    print(site)
    if site=='Jingdong':
        break
else:
    print('traversal completed')
```

The following information is output in the terminal.

```
Baidu
Google
Jingdong
```

Since a break occurred when traversing to "Jingdong", the terminal was traversed and "Taobao" and else clause were not output.

3. while statement

The while statement contains both a condition and a loop statement, and is used to determine whether to execute the code block based on the result of the condition. It is used as follows:

```
while condition:
    statements
else:
    statements
```

The while statement can be used alone or combined with an else clause. It is the same as the if statement. The statement in the else will be executed only when the condition is False.

```
a=3
while a:
    print(a)
    a-=1
else:
    print('traversal completed')
```

The following information is output in the terminal.

```
3
2
1
traversal completed
```

When a decreases to 0, the condition of while is False, so the else clause is executed.

4. break and continue

The break statement is similar to the statement in C language and is used to jump out of the latest for or while loop.

The continue statement is also borrowed from the C language and means to

Chapter1　Maze Robot Simulation Foundation | 029

continue executing the next iteration of the loop.

Comparing the execution results shown in Table 1-1-2, we can find the following results:

break terminates the entire loop, and the remaining loops are no longer executed;

continue ends this loop early, the following code is no longer executed, and it directly enters the next loop.

Table 1-1-2　Comparison table of execution results

break	continue
```\nn = 5\nwhile n:\n n -= 1\n if n == 2:\n break\n print(n)\nprint('end of loop.')\n```	```\nn = 5\nwhile n:\n n -= 1\n if n == 2:\n continue\n print(n)\nprint('end of loop.')\n```
```\n4\n3\nend of loop.\n```	```\n4\n3\n1\n0\nend of loop.\n```

Task 2　Master Python Data Types

This task is to learn basic Python data types and master common methods.

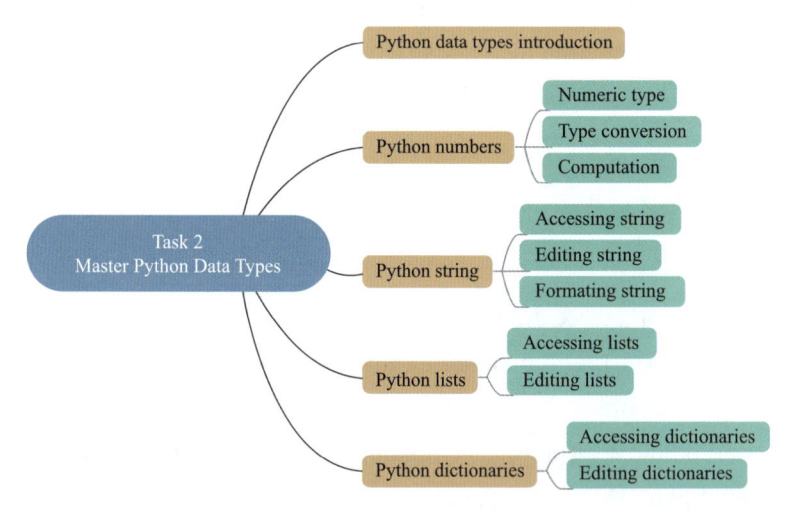

Ⅰ. Python data types introduction

Different from other languages, data types in Python do not refer to variables.

Variables in Python do not need to be declared in advance. They must be assigned a value before use. The variable will not be actually created until it is assigned.

In Python, a variable is a variable, it has no type. What we call "type" is the type of the object in the memory address pointed to by the variable.

The equal sign (=) is used to assign a value to a variable;

The left side of the equal sign (=) operator is a variable name, which can be any name that conforms to Python syntax, even Chinese;

The right side of the equal sign (=) operator is the value stored in the variable.

There are many data types in Python. Commonly used basic types are number, string, bool, list, tuple and dictionary.

bool is relatively simple, only containing True and False; tuple can be considered as the immutable form of list, and its basic usage is similar to list, so I won't explain it too much. The following mainly explains the other four basic types.

Ⅱ. Python numbers

Python numbers are used to store numerical values.

Python numbers are immutable data types, which means that if you change the value of the number, the memory space will be reallocated.

```
>>> var1=66
>>> var2=88
>>> var1, var2
    (66, 88)
```

1. Numeric type

Python supports three different numeric types:

Integer type (int): Often called an integer, is a positive or negative integer without a decimal point.

Floating point type (float): The floating point type consists of an integer part and a decimal part. The floating point type can also be expressed using scientific

notation ($2.5e2=2.5\times10^2=250$).

Complex number (complex): A complex number consists of a real part and an imaginary part, which can be represented by a+bj, or complex(a, b). The real part a and the imaginary part b of the complex number are both floating point types.

2. Type conversion

Different numerical types can be converted using the corresponding functions:

int(x) converts x to an integer.

float(x) converts x to a floating point number.

complex(x) converts x to a complex number, with the real part being x and the imaginary part being 0.

complex(x, y) converts x and y into a complex number, with the real part being x and the imaginary part being y. x and y are numeric expressions.

3. Computation

Python numbers can also be directly used for mathematical operations.

Assume that variable a=5 and variable b=2, as shown in Table 1-1-3.

Table 1-1-3　Operator comparison table

Operator	Description	Example
+	Add, add two objects	a+b output result 7
−	Subtract, get a negative number or subtract one number from another number	a−b output result 3
*	Multiply, multiply two numbers or return a string repeated several times	a*b output result 10
/	Division, x divided by y	b/a output result 0. 4
%	Takes remainder and returns the remainder of division	b%a output result 2
**	Power, returns the y power of x	a**b is 5 raised to the power 2, 25

//	Divide by integers and take integers in the smaller direction	>>> 9//2
		4
		>>> −9//2
		−5

Python also provides a large number of mathematical functions and trigonometric functions. Please check the official website documentation for specific instructions.

Ⅲ. Python string

Python string (string, abbreviated as str) , used to store strings, is the most commonly used data type in Python. Strings can be created using ' ' or " ".

Python strings are immutable data types, which means that if you change the value of the string, the memory space will be reallocated.

```
>>> var1 = 'Hello TQD!'
>>> var2 = 'Hello Python!'
>>> var1, var2
('Hello TQD!', 'Hello Python!')
```

1. Accessing string

Python strings support the following operations, as shown in Table 1-1-4.

Table 1-1-4   Access string

String operation	Description	Example	
+	Connector (same type)	>>> 'he'+'llo' 'hello'	
*	Repeat operation	>>> 'hello'*3 'hellohellohello'	
[i]	Index. The subscript starts from 0, and negative numbers mean from right to left	>>> 'hello'[1] 'e'	>>> 'hello'[-1] 'o'

[start:end:step]	Slice, the first ':' cannot be omitted, the others can be omitted. All three parameters can be negative, indicating from right to left	>>> 'hello'[:-1]
		'hell'
		>>> 'hello'[::-1]
		'olleh'	

in	in member operator True/False	>>> 'h' in 'hello' True	
not in	Non-member operator True/False	>>> 'a' not in 'hello' True	
del	Delete a string. str is immutable, so you cannot delete indexes or slices, you can only delete the entire str	>>> s='hello' >>> del s	

for...in...	Iteration	>>> s = 'word'
		>>> for i in s:
		        print(i)
		w
		o
		r
		d	

2. Editing string

Since Python strings are immutable, editing a string means indexing, slicing, etc. the original string, and then concatenating other strings to generate a new string. Commonly used editing methods are shown below:

(1) Case conversion

```
str. capitalize()-> str
```

Capitalize first letter.

```
>>> 'hello python'. capitalize()
'Hello python'
str. swapcase()-> str
```

Swap case.

```
>>> 'Hello Python'. swapcase()
'hELLO pYTHON'
str. lower()-> str
```

All lowercase.

```
>>> 'HELLO PYTHON'. lower()
'hello python'
str. upper()-> str
```

All uppercase.

```
>>> 'hello python'. upper()
'HELLO PYTHON'
str. title()-> str
```

Title case, the first letter of each word needs to be capitalized.

```
>>> 'hello python'. title()
'Hello Python'
```

(2) Find and replace

```
str. find(sub[, start[, end]])) ->int
```

Search for sub and return the index; if sub does not exist, return -1.

```
>>> 'hello python'. find('o')
4
>>> 'hello python'. find('a')
```

```
-1
str. rfind(sub[, start[, end]]) ) -> int
```

Find sub from the right and return the index.

```
>>> 'hello python'. rfind('o')
10
str. index(sub[, start[, end]) -> int
```

Find sub and return the index. The difference from find is that if sub does not exist, index will report an error.

```
>>> 'hello python'. index('o')
4
>>> 'hello python'. index('a')
Traceback (most recent call last):
  File "<pyshell#10>", line 1, in <module>
    'hello python'. index('a')
ValueError: substring not found
str. rindex(sub[, start[, end]])) -> int
```

Find sub and return the index.

```
>>> 'hello python'. rindex('o')
10
str. count(sub, start=None, end=None) -> int
```

Count the number of times sub appears.

```
>>> 'hello python'. count('l')
2
str. replace(old, new [, max]) ->str
```

Replace character.

```
>>> 'hello python'. replace('p', 'P')
'hello Python'
```

(3) Fill and align

```
str. center(width, fillchar=' ') ->str
```

Centered.

```
>>> 'TQD'. center(11)
'TQD'
>>> 'TQD'. center(11, '~')
```

Chapter1 Maze Robot Simulation Foundation | 035

```
'~~~~TQD~~~~'
str. ljust(width, fillchar=' ') ->str
```

Align left.

```
>>> 'TQD'. ljust(11)
'TQD'
str. rjust(width, fillchar=' ') ->str
```

Align right.

```
>>> 'TQD'. rjust(11)
'TQD'
```

(4) Remove and match

```
str. strip([chars]) ->str
```

Remove leading and trailing spaces or characters belonging to chars.

```
>>> 'Hello python'. strip()
'Hello python'
str. lstrip([chars]) ->str
```

Remove leading spaces or characters belonging to chars.

```
>>> 'Hello python'. lstrip()
'Hello python'
str. rstrip([chars]) ->str
```

Remove trailing spaces or characters belonging to chars.

```
>>> 'Hello python'. rstrip()
'Hello python'
```

(5) Connection and division

```
str. join(iterable) ->str
```

Concatenates all elements of iterable, separated by str.

```
>>> ''. join(('Hello', 'TQD'))
'HelloTQD'
>>> ','. join(('Hello', 'TQD'))
'Hello, TQD'
str. split(sep=None, maxsplit=-1) ->list
```

Split the string by sep.

```
>>> 'Hello Python! Hello World!'. split(' ')
['Hello', 'Python!', 'Hello', 'World!']
str. rsplit(sep=None, maxsplit=-1)    ->list
```

Split the string by sep, from right to left.

```
>>> 'Hello Python! Hello World!'. rsplit(' ', 1)
['Hello Python! Hello', 'World!']
```

3. Formating string

When outputting a string in a program, it is often necessary to adjust it according to the actual situation, such as the type, time, quantity, name, age, etc. in the string.

It is a very tedious task to create a one-to-one corresponding variable for each string that needs to be output. At this time, you can consider using formatted strings.

```
'the Python course will start at 10:00, with a total of 22
students participating.'
    'the Math course will start at 11:00, with a total of 30
students participating.'
    'the English course will start at 14:00, with a total of 28
students participating.'
```

First create a template string, and when using it, format and output it according to the actual situation.

```
'the {} course will start at {}, with a total of {} students participating.'
```

In Python, the curly brackets "{}" in a string can be used as placeholders, indicating that this is a variable and can be formatted. There are three ways to format:

% formatting: This is the formatting method used in the Python2 era;

format formatting: The format() method of string has a very high degree of freedom;

f-string formatting: f-string is a string marked with 'f' or 'F' prefix. You can directly use the variables that appeared above in "{}".

Since % formatting is no longer recommended, the following mainly introduces format formatting and f-string formatting.

1) format formatting

```
str. format(*args, **kwargs)
```

Chapter1　Maze Robot Simulation Foundation | 037

(1) sequence formatting

Replace the '{}' placeholder field in str with numerical sequence.

```
>>> '{0} {2}{1}'. format('hello', 'ld', 'wor')
'hello world'
```

If the sequence number is in the order of 0, 1, 2, ... , it can be omitted, but in this case the number of '{}' and format parameters must be the same.

```
>>> '{} {}'.format('hello', 'world')
hello world
>>> template ='the {} course will start at {}, with a total of {}
students participating.'
>>> template.format('Python', '10: 00', 22)
'the Python course will start at 10:00, with a total of 22 students
participating.'
>>> template.format('数学', '11: 00', 30)
'the Math course will start at 11:00, with a total of 30 students
participating.'
>>> template.format('英语', '14: 00', 28)
```

(2) Keyword formatting

Replace '{}' placeholder fields according to keywords.

```
>>> template ='the Python course will start at 10:00, with a
total of 22 students participating.'
>>> template.format(subject='Python', time='10: 00', students=22)
'the Python course will start at 10:00, with a total of 22
students participating.'
>>> template.format(subject='数学', time='11: 00', students=30)
'the Math course will start at 11:00, with a total of 30
students participating.'
>>> template.format(subject='英语', time='14: 00', students=28)
'the English course will start at 14:00, with a total of 28
students participating.'
```

(3) Number formatting

Format the numbers in a string for display.

```
>>> '{:.2f}'. format(3.1415926)      # keep two decimal places
'3.14'
>>> '{:>5}'. format(3.14)            # right align, total width
                                     # of 5 characters
' 3.14'
>>> '{:~>5}'. format(3.14)     # right align, total width of 5
                              # characters, filled with ~
```

```
'~3.14'
>>> '{:>5.2f}'. format(3.1415926)
# right align, total width of 5 characters; keep two decimal places
' 3.14'
```

2) f-string formatting

Compared to format, it allows you to directly execute expressions inside '{}' braces.

```
>>> f'{3>2}'
'True'
>>> food = ['watermelon', 'grape', 'banana']
>>> f'{food[0]}'
'watermelon'
>>> class Car:
    color = 'red'
>>> car = Car()
>>> f'{car.color}'
'red'
```

f-strings also support formatting numbers within the string.

```
>>> value = 3.1415926
>>> f'{value:.2f}'
'3.14'
>>> f'{value:>5.2f}'
' 3.14'
>>> f'{value:~>5.2f}'
'~3.14'
```

IV. Python lists

Python lists are sequences used to store multiple data items. Lists can contain elements of any type, and they can be homogeneous or heterogeneous. You can create lists using square brackets "[]".

Python lists are mutable data types, allowing you to add, delete, or modify their elements.

```
>>> list1 = ['C/C++', 'Java', 'Python', 'Go']
>>> list2 = ['xiao ming', 20, 88]
```

1. Accessing lists

Python lists support the following operations, as shown in Table 1-1-5. Create a variable: >>> l = [22,33,44,55].

Chapter1　Maze Robot Simulation Foundation | 039

Table 1-1-5　Accessing list

List operation	Description	Example	
+	Connector (same type)	>>> [1,2,3]+[7,8] [1, 2, 3, 7, 8]	
*	Repeat operation	>>> [1,2]*3 [1, 2, 1, 2, 1, 2]	
[i]	Index The subscript starts from 0, and negative numbers mean from right to left	>>> l[0] 22	>>> l[-1] 44

[start:end:step]	Slice, the first ':' cannot be omitted, the others can be omitted. All three parameters can be negative, indicating from right to left	>>> l[:2]
		[22, 33]
		>>> l[::2]
		[22, 44]	

in	In member operator True/False	>>> 66 in l False	
not in	Non-member operator True/False	>>> 44 not in l True	

del	Delete elements or slice del list[i]
	Delete the entire list del list
		>>> del l[2]
		>>> l
		[22, 33, 55]	

for...in...	Iteration	>>> for i in l: 　　　print(i) 22 33 44 55	

2. Editing list

Python lists are mutable types of data, and adding, deleting, or modifying them modifies the list itself.

(1) Searching and counting

```
list. index(x[, start[, end]]) ->int
```

To find the index of x.

```
>>> list1 = ['C/C++', 'Java', 'Python', 'Go']
>>> list1. index('Python')
2
list. count(x) ->int
```

Count the number of times a certain element appears in the list.

```
>>> list1. count('Python')
1
```

(2) Insertion and extension

```
list. append(x) ->None
```

Append, add element x to the end of the list.

```
>>> list1. append('C#')
>>> list1
['C/C++', 'Java', 'Python', 'Go', 'C#']
list. insert(i, x) ->None
```

Insert, insert element x at index i.

```
>>> list1. insert(3, 'JavaScript')
>>> list1
['C/C++', 'Java', 'Python', 'JavaScript', 'Go', 'C#']
list. extend(iterable) ->None
```

Extend, append the iterable to the end of the list.

```
>>> list1. extend(('VB', 'SQL'))
>>> list1
['C/C++', 'Java', 'Python', 'JavaScript', 'Go', 'C#', 'VB', 'SQL']
```

(3) Delete and clear

```
list. pop([i]) ->object
```

Delete elements based on an index.

```
>>> list1. pop()
'SQL'
>>> list1. pop(-2)
'C#'
list. remove(x) ->None
```

Delete elements based on their value, removing only the first occurrence.

```
>>> list1. remove('Go')
>>> list1
['C/C++', 'Java', 'Python', 'JavaScript', 'VB']
list. clear() ->None
```

Clear the list.

```
>>> list1. clear()
>>> list1
[]
```

(4) Sort

```
list. sort(*, key=None, reverse=False) ->None
```

Sorting a list.

Chapter1　Maze Robot Simulation Foundation | 041

```
>>> list1 = ['C/C++', 'Java', 'Python', 'Go']
>>> list1. sort()
>>> list1
['C/C++', 'Go', 'Java', 'Python']
list. reverse() ->None
```

Reverse the elements in the list, regardless of their size.

```
>>> list1. reverse()
>>> list1
['Python', 'Java', 'Go', 'C/C++']
```

V. Python dictionaries

Python dictionaries (dict) are also part of sequences and are containers for key-value pairs. You can create dictionaries using curly braces {}.

Python dictionaries are mutable data types, which means you can add, delete, or modify key-value pairs within them.

```
>>> dict1 = {'name':'li', 'age':18, 'score':90}
```

1. Accessing dictionaries

Each element in a Python dictionary is a key-value pair and supports the following operations, as shown in Table 1-1-6. Create a variable, >>> d = {'a':1, 'b':2}.

Table 1-1-6　Access dictionary

Dictionary operation	Description	Example

[key]	Access or assign values by key, dict does not support indexing and slicing	>>> d['a'] = 5
		 >>> d['a']
		 5

in	In member (key) operator True/False	>>> 'a' in d True
not in	Not in non-member (key) operator True/False	>>> 'd' not in d True

del	Deletes an element del dict[key]
	 Delete the entire dictionary del dict
		>>> del d['a']
		 >>> d
		 {'b': 2}

for...in...	Iteration	>>> for i in d: 　　print(i) a b

Different from the strings and lists, dictionaries do not support indexing and slicing.

2. Editing dictionaries

Python dictionaries are mutable type data. Adding, deleting or modifying is modifying the dictionary itself.

(1) Find and count

```
dict. get(key[, default]) ->value
```

Returns the value of key, or default if it does not exist.

```
>>> dict1 = {'name':'li', 'age':18, 'score':90}
>>> dict1. get('name')
'li'
dict. setdefault(key[, default]) ->value
```

Returns the value of key. If it does not exist, add the key:default key-value pair and return default.

```
>>> dict1. setdefault('subject', 'Python')
'Python'
>>> dict1
{'name': 'li', 'age': 18, 'score': 90, 'subject': 'Python'}
dict. items() -> dict_items[_KT, _VT]
```

Returns a view composed of key-value pairs (tuples) , which can be iterated, but slices cannot be indexed. If necessary, list expansion can be used.

```
>>> dict1. items()
dict_items([('name', 'li'), ('age', 18), ('score', 90),
('subject', 'Python') ])
dict. keys() -> dict_keys[_KT, _VT]
```

Returns a view composed of keys, which can be iterated, but slices cannot be indexed. If necessary, list expansion can be used.

```
>>> dict1.keys()
dict_keys(['name', 'age', 'score', 'subject'])
dict. values() -> dict_values[_KT, _VT]
```

Returns a view composed of values, which can be iterated, but cannot index slices. If necessary, list expansion can be used.

```
>>> dict1.values()
dict_values(['li', 18, 90, 'Python'])
```

Chapter1　Maze Robot Simulation Foundation ｜ 043

(2) Delete and clear

```
dict.pop(key[,default]) ->value
```

Delete a key-value pair and return the value. If the key doesn't exist and no default value is provided, it will raise a KeyError.

```
>>> dict1.pop('age')
18
>>> dict1
{'name': 'li', 'score': 90, 'subject': 'Python'}
dict.popitem() ->tuple
```

Delete the last key-value pair and return it as a tuple.

```
>>> dict1.popitem()
('subject','Python')
>>> dict1
{'name': 'li', 'score': 90}
dict.clear() ->None
```

Clear all elements from the dictionary.

```
>>> dict1.clear()
>>> dict1
{}
```

Task 3　Learn Python Functions and Classes

This task is to learn Python functions and classes and master commonly used methods.

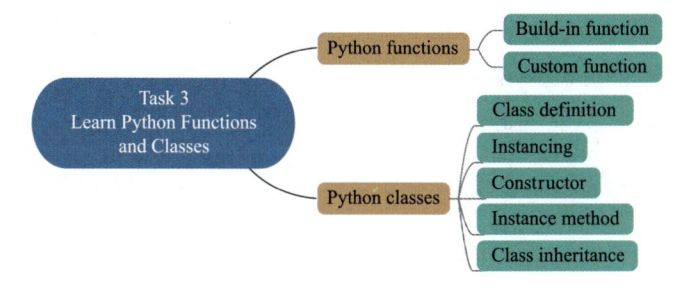

Ⅰ. Python functions

A function is an encapsulated and reusable code segment used to implement

a single or related set of functionalities. Functions enhance the modularity of a program and the reusability of code.

1. Built-in function

In Python, the interpreter includes many built-in functions that can be used directly without the need for importing, as shown in Table 1-1-7. For detailed usage instructions for each built-in function, please refer to the official documentation on the Python website.

Table 1-1-7　Built-in function comparison table

A	E	L	R
abs()	enumerate()	len()	range()
aiter()	eval()	list()	repr()
all()	exec()	locals()	reversed()
any()			round()
anext()	F	M	
ascii()	filter()	map()	S
	float()	max()	set()
B	format()	memoryview()	setattr()
bin()	frozenset()	min()	slice()
bool()			sorted()
breakpoint()	G	N	staticmethod()
bytearray()	getattr()	next()	str()
bytes()	globals()		sum()
		O	super()
C	H	object()	
callable()	hasattr()	oct()	T
chr()	hash()	open()	tuple()
classmethod()	help()	ord()	type()
compile()	hex()		
complex()		P	V
	I	pow()	vars()
D	id()	print()	
delattr()	input()	property()	Z
dict()	int()		zip()
dir()	isinstance()		
divmod()	issubclass()		_
	iter()		__import__()

2. Custom function

In addition to using built-in functions, you can also customize functions to implement the functions you need.

(1) Function declaration

```
def function name ([parameter]):
    # internal code
    return expression
```

Python uses the keyword "def" to define functions;

The function name is mandatory and can be any name that follows Python

syntax, even in Chinese;

After the function name, parentheses must be used to define parameters; if no parameters are needed, they can be omitted;

A colon ':' must follow the parentheses, indicating the beginning of the code block;

The first-level indentation inside the function must be consistent, indicating that this code block is a whole;

At the end of the function, you should use 'return' to return a value; if there is no 'return' statement, Python will default to return None.

```
def function():
    return 1
print(function()) # Output the return value, which is 1
```

If a function needs to use parameters, you can define them inside the parentheses after the function name. The parameter names here serve as variable names. There are primarily two types of parameters: positional parameters and keyword parameters.

(2) Positional parameters

Positional parameters are also called required parameters, Sequential parameters are the most important and must be provided explicitly when calling the function. Positional parameters must be passed in order, one-to-one correspondence, no more, no less.

Note: Python does not perform data type checking when passing function parameters, theoretically allowing you to pass any data type. However, during actual computation, if a data type error is encountered, it will raise an exception.

```
def fun(a, b, c):
    return a + b + c


print(fun(5, 2, 1))      # 8
```

(3) Keyword parameters

When calling a function, all parameters are passed into the function by position (order). If the order of passing does not match the order of use, it can lead to various issues, including crashes.

Using keyword parameters when passing parameters can indeed help you

avoid issues related to parameter order, as you don't need to follow a specific sequence. Keyword parameters involve passing parameters in the format "parameter name=value," allowing you to specify which argument corresponds to which parameter by name.

```
def fun(a, b):
    return a ** b


print(fun(2, 3))                # 8
print(fun(3, 2))                # 9
print(fun(b=3, a=2))            # parameters are passed by key word,
                                  regardless of the order,8
```

(4) Default parameters

When defining a function, you can indeed provide a default value for a parameter. These parameters with default values are also considered keyword parameters. When calling the function, you have the option to pass a custom value for the keyword parameter or omit it, in which case the default value will be used.

```
def fun(a=2, b=3):
    return a ** b

print(fun())                    # 8
print(fun(5, 2))                # 25
```

When combining positional parameters and keyword parameters in Python, the syntax requires that keyword parameters must come after positional parameters; otherwise, it will trigger an exception.

```
def fun(a, b=3):
    return a ** b


print(fun(5))                   # 125
```

Placing keyword parameters before positional parameters will indeed trigger an exception.

```
def fun(a=3, b):
    return a ** b
SyntaxError: non-default argument follows default argument
```

II . Python classes

Python is an object-oriented language, and classes in Python provide all the

basic functions of object-oriented programming. The class inheritance mechanism allows a derived class to override any method in the base class, and the method can call the method of the same name in the base class.

1. Class definition

```
class class name:
        def function name(self, [parameter]):
                # internal code
                return expression
   ...
```

Python uses the keyword "class" to define classes;

A class name must be provided, and it's recommended to start with an uppercase letter;

If the class inherits from other classes, you can add parentheses after the class name and place the base class inside them;

Within the class, you can define functions, called instance methods, which must accept a parameter "self" to represent the instance object itself;

Variables defined outside functions are class variables and are similar to regular variables;

Variables defined within functions are instance attributes and must be represented as "self. variable".

```
class Car:

    def __init__(self, person, color, seats):
        self. person = person
        self. color = color
        self. seats = seats

    def say(self):
        a = 2
    print(f'{self.person}'s car is {self.color}, with a total of
{self.seats} seats')

    def get_color(self):
        return self. color

    def set_color(self, color):
        self. color = color
```

2. Instancing

As shown in the above example, Car is a class, pointing to the memory

address; and Car() is execution, an instantiation of the class. If parameters are used, they need to be passed in during instantiation.

```
mycar = Car('xiao ming', 'red', '4')
```

3. Constructor

Constructor is a special instance method in a class. Its name is fixed as "__init__", and it is executed first when a class is instantiated. You can think of it as the initialization method. Any parameters passed during instantiation will be received by the constructor method and used for initializing the instance.

```
def __init__(self, …):
    code block
```

All instance methods, including the constructor, must accept a "self" parameter, which represents the instance object itself. Instance attributes must be represented in the form of "self,…"

4. Instance method

Instance methods are functional methods within a class that can be used both internally and exposed externally. If an instance method doesn't fully meet the requirements, you can consider inheriting the class and overriding the method

```
def name(self,…):
    code block
```

In addition to the self parameter, instance methods can add other additional parameters. But please note that these additional parameters are not passed in during instantiation, so when calling the instance method, additional parameters need to be passed in.

5. Class inheritance

The inheritance mechanism is often used to create new classes with similar functions to existing classes, or the new class only needs to add or modify some properties or methods based on the existing class, but does not want to directly copy the existing class code to the new class. In other words, by using inheritance as a mechanism, classes can be easily reused.

```
class Parent:
    pass
```

```
class Child(Parent):
    pass
```

① Child classes can directly access and use attributes and methods of the parent class.

② Child classes can define new attributes and methods.

③ Child classes can override (redefine) attributes and methods that were originally in the parent class.

Task 4　Master Python Extension Libraries

This task involves understanding Python extension libraries and becoming proficient in using both common standard libraries and third-party libraries.

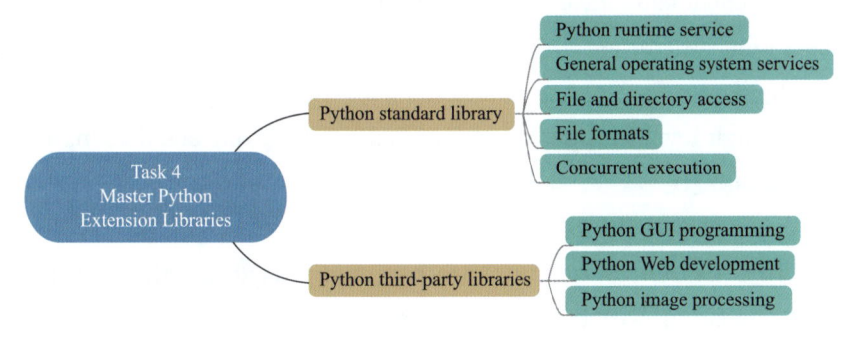

Ⅰ. Python standard library

The standard library in Python consists of integrated extension libraries that are installed locally along with the Python client. As there are more than 240 modules in the Python standard library, this book only introduces some commonly used standard libraries.

1. Python runtime service

Sys　provides access to system-related parameters and functions. These variables may be used by the interpreter or provided by the interpreter itself. These functions can affect the behavior of the interpreter.

2. General operating system services

(1) os

Offers various operating system interface functions. If you need to read or

write a file, you can use open(). If you want to work with file paths, you can refer to the os. path module. If you need to read all lines from all files specified on the command line, you can check out the fileinput module. For creating temporary files and directories, you can use the tempfile module. For more advanced file and directory handling, consider using the shutil module.

(2) time

Provides access to time-related functions and conversions. For more advanced time and date operations, you can also refer to the datetime and calendar modules.

3. File and directory access

(1) os. path

Provides common path manipulation functions. Different operating systems have different path name conventions, so the os. path standard library integrates conversion functions to ensure that the output paths are always suitable for the local operating system.

(2) shutil

Offers high-level file operations, especially for copying and deleting files.

4. File formats

(1) configparser

INI files (initialization files) are commonly used initialization files on Windows, governing various Windows configurations. The configparser module provides functionality for parsing INI files.

(2) csv

csv (comma separated values) format is one of the most common input and output file formats for spreadsheets and databases. The csv module provides reading and writing capabilities for CSV files.

5. Concurrent execution

(1) threading

Provides thread-based parallelism. Due to the presence of the global interpreter lock (GIL) , only one thread can execute at a time (although certain performance-oriented libraries may release this limitation). Therefore, if you want to fully utilize multi-core computers, it's recommended to use multiprocessing. If the tasks you're running concurrently are I/O-bound, then multithreading can still be a suitable choice.

(2) multiprocessing

Provides process-based parallelism. It effectively bypasses the GIL by using subprocesses rather than threads. Therefore, the multiprocessing module allows programmers to fully utilize the multi-core performance of a computer.

(3) subprocess

Manages subprocesses and allows you to connect their input, output, and error pipes, as well as retrieve their return values.

Ⅱ. Python third-party libraries

Third-party libraries in Python are extensions that are not integrated into the Python client by default. They can be installed and used using the pip command.

1. Python GUI programming

GUI, which stands for graphical user interface, allows users to interact with software through graphical elements such as buttons, menus, and windows. Python provides the built-in tkinter module for quickly developing simple desktop applications. Additionally, third-party libraries such as PyQt, PySide, PySimpleGUI, Kivy, wxPython, and more can be used for GUI programming. Here, we recommend using PySide.

PySide is a Python binding for the cross-platform application framework Qt. Qt is a well-regarded cross-platform C++ framework for developing graphical user interfaces. PySide is maintained by the Qt Company, and the latest version is PySide6. It allows users to develop large and complex GUI applications using Qt within the Python environment.

PySide6 supports the LGPL license, enabling the development of closed-source programs through dynamic linking. Applications can be released in any form (commercial, non-commercial, open-source, non-open-source, etc.).

```
pip install PySide6
```

As shown below, this is a very basic button program.

```
from PySide6. QtWidgets import QWidget, QApplication, QPushButton
import sys

class Mydemo(QWidget):
    def __init__(self):
        super()    . __init__()
        btn = QPushButton('button', self)
```

```
            btn. move(50, 50)
            btn. setStyleSheet("QPushButton:pressed {color:blue}")

if __name__ == '__main__':
    app = QApplication(sys. argv)
    win = Mydemo()
    win. show()
    sys. exit(app. exec())
```

After running the program, you will see the GUI interface, and when you click the button, the text will turn blue, as shown in Figure 1-1-10.

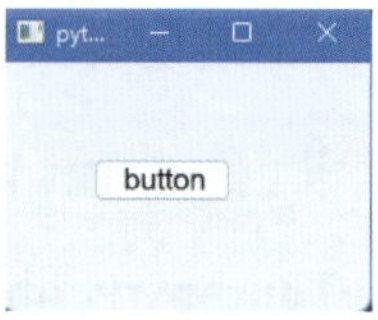

2. Python Web development

Typical Web application development frameworks include Django, Flask, and Pyramid. Students can choose to learn software based on their own interests. At this time, we recommend using Flask for simple Web development.

Figure 1-1-10　Basic GUI interface

Flask is a lightweight Web application framework written in Python. It's often referred to as a microframework. The "micro" in Flask doesn't mean that you have to put your entire Web application in a single Python file; rather, it means that Flask is designed to keep your code simple and easy to extend. Flask's core components are straightforward, but it offers strong extensibility and compatibility. Programmers can use the Python language to quickly implement a website or Web service, and users can extend it as needed.

To install Flask, you can use the following command:

```
pip install flask
```

As shown below, this is a very simple Web program.

```
from flask import Flask

app = Flask(__name__)
@app. route('/')
def hello_world():
    return 'Hello World!'
if __name__ == '__main__':
    app. run()
```

After running it, you can access the content in a web browser by visiting http://127. 0. 0. 1:5000, as shown in Figure 1-1-11.

3. Python image processing

In Python, image processing is often a necessary task, including tasks like

reading, writing, and recognition. There are two commonly used libraries for this: OpenCV and Pillow. Here, we recommend using OpenCV.

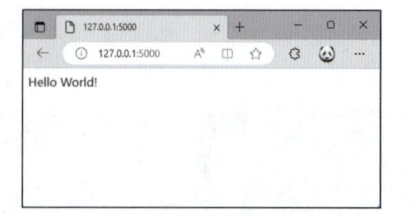

OpenCV is an open-source computer vision library sponsored by Intel. It can

Figure 1-1-11   Basic Web service

run on Linux, Windows, Android, and MacOS operating systems and provides a Python interface, making it very convenient for implementing various general-purpose algorithms in image processing and computer vision.

You can install OpenCV using the following command:

```
pip install opencv-python
```

As shown below, this is a very simple color picker program.

```python
import cv2

def mouseColor(event, x, y, flags, param):
    if event == cv2. EVENT_LBUTTONDOWN:
        print('HSV:', hsv[y, x])

img = cv2. imread('2. jpg')
hsv = cv2. cvtColor(img, cv2. COLOR_BGR2HSV)
cv2. imshow("Color Picker", img)
cv2. setMouseCallback("Color Picker", mouseColor)
if cv2. waitKey(0)== ord('q'):
    cv2. destroyAllWindows()
```

After running the program, you can click on any position on the image to obtain the HSV components of that location, as shown in Figure 1-1-12.

Figure 1-1-12   Color picker program

Questions and Summary

1. Different projects require different extension libraries. How to maintain multiple projects?

2. What other common methods are used in pip?

3. In order to improve the efficiency of project code, it is recommended to write specific code as a function or class and achieve reuse through calls.

Project 2

Learning Environment Construction

Learning objectives

1. Master how to install and use the Ubuntu system.

2. Master the installation of ROS2 and understand the development of ROS2.

3. Master the installation of Gazebo and integrate it with ROS2.

Task 1 Getting Started with the Ubuntu System

The purpose of this task is to have a certain understanding of the Ubuntu system, learn how to install the Ubuntu system, and be familiar with common system commands.

I. Introduction to Ubuntu system

1. Linux system introduction

The Linux system was developed from the kernel released by Linus Torvalds in the newsgroup in 1991. Because it was free and freely disseminated

Chapter1　Maze Robot Simulation Foundation | 055

at the beginning of its release, it supports multi-users, multi-tasking and multi-threading, and compatible with the POSIX standard, which made it support some tools and software running the mainstream system Unix at that time, which attracted many users and developers, and gradually developed and expanded to this day.

Since the Linux kernel itself is open source, a complete operating system is formed based on the Linux kernel with various system management software or application tool software. As shown in Figure 1-2-1, a complete Linux system is like a car, the Linux kernel constitutes the most critical engine, and different distribution versions are like different car models using the same engine.

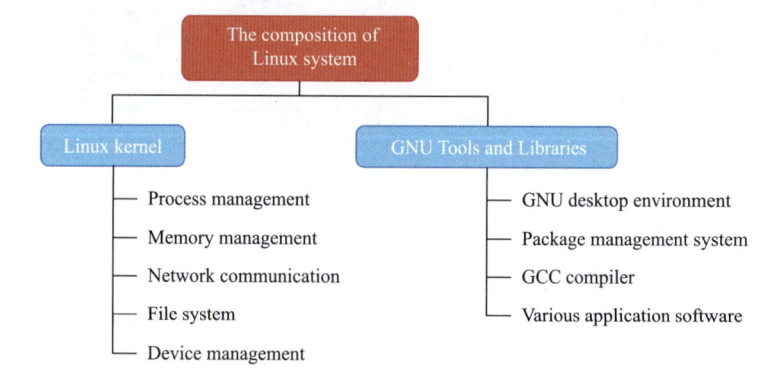

Figure 1-2-1　The composition of Linux system

People create Linux distributions for specific purposes. They focus on different aspects, resulting in a wide variety of Linux distributions. Among them, Debian, Suse, and Fedora distributions are the most common, as shown in Figure 1-2-2.

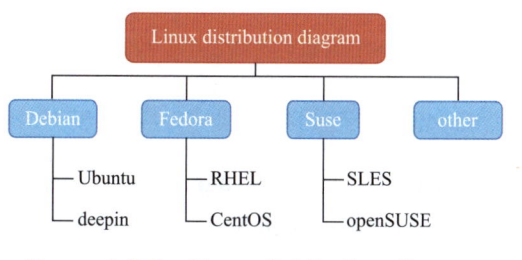

Figure 1-2-2　Linux distribution diagram

For users trying Linux for the first time, we recommended to just understand that these are all Linux systems. Moreover, our development host will be using Ubuntu.

2. Ubuntu system introduction

Ubuntu is a Linux distribution derived from Debian. The Ubuntu 22.04 desktop system is shown in Figure 1-2-3, please search the URL in the search engine by yourself.

Figure 1-2-3   Ubuntu 22.04 desktop

Ubuntu focuses on users and usability, and has the following characteristics:

(1) Complete desktop system

Ubuntu provides everything you need to run your organization, school, home or business. All necessary applications such as office software, browsers, e-mail and multimedia applications, etc. are preinstalled. The Ubuntu software center offers thousands of games and applications.

(2) Open source

Ubuntu has always been free to download, use and share.

(3) Security

Ubuntu is one of the most secure operating systems, with a built-in firewall and virus protection software.

(4) Accessible

Computing is used for all people, regardless of nationality, gender, etc. Ubuntu is fully translated into more than 50 languages and includes the necessary assistive technologies.

(5) System interface effect

Ubuntu supports high-definition touchscreens, desktop UI scaling, and touchpad gestures. 22.04 LTS updates the iconic Yaru theme with system-wide dark style preference support, accent colors and the largest collection of

Chapter1　Maze Robot Simulation Foundation

community wallpapers to date.

(6) Broad hardware support

Canonical works closely with Dell, Lenovo and HP to demonstrate that Ubuntu is available on the widest range of laptops and workstations. Can provide an unprecedented seamless Ubuntu experience, and more hardware choices than ever before.

Ubuntu is not only suitable for desktop computers, it is also widely used in data centers around the world, providing support for a variety of servers, and it is also the most popular operating system in cloud computing.

The Ubuntu file directory is shown in Figure 1-2-4.

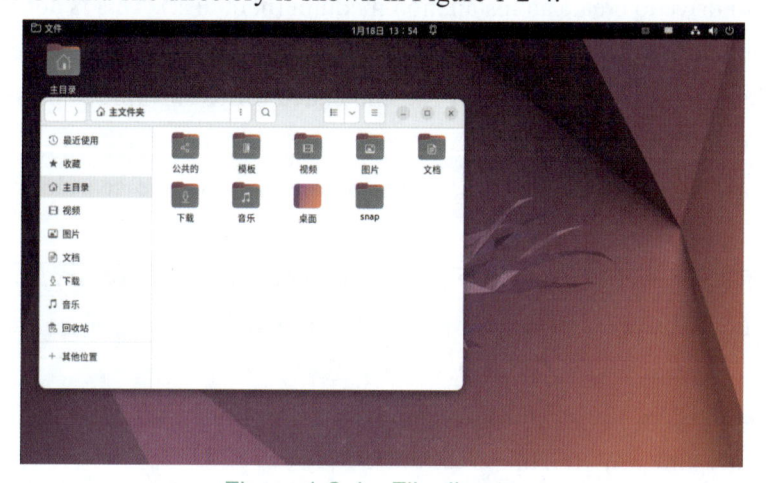

Figure 1-2-4　File directory

The Ubuntu system settings are shown in Figure 1-2-5.

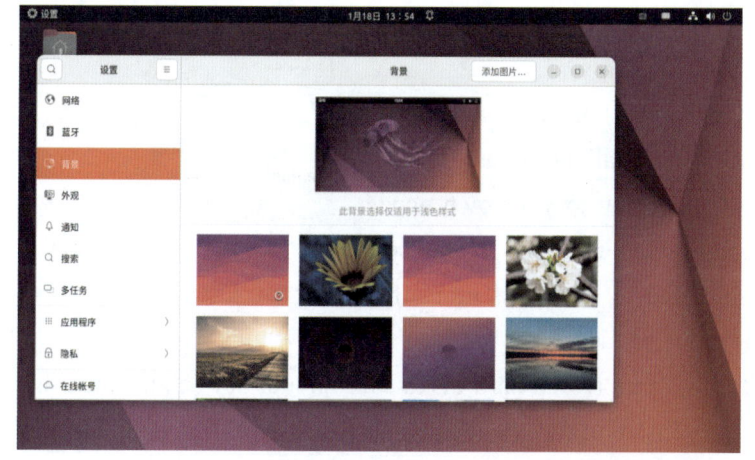

Figure 1-2-5　System settings

II. Ubuntu system installation

This book is based on the latest long-term supported version of Ubuntu, Ubuntu 22.04 LTS. It is recommended to install the Ubuntu system directly on a physical machine. If it is installed in a virtual machine, the performance of the simulation may be affected.

1. Preparation

First download the Ubuntu image file, please search for the download address in the search engine.

To install Ubuntu Desktop, you also need to write the downloaded ISO to a USB flash drive to create an installation medium (boot disk). This is not the same as copying an ISO and requires some custom software. In this book, based on the Windows system, use the officially recommended Rufus software to make a USB boot disk.

Please search for the Rufus download address in the search engine by yourself.

Rufus software does not need to be installed after downloading, and it can be used immediately with Windows 7 or higher (32/64 bit).

Then make a USB boot disk, the Rufus settings are shown in Figure 1-2-6, click Start, and wait for the creation to complete.

Then compress the hard disk. If you directly overwrite the Windows system, this step is unnecessary.

Compress the hard disk and divide the hard disk for installing Ubuntu. The size can be set according to your own needs. Ubuntu itself takes up about 10 GB, so it is recommended to compress more than 30 GB, as shown in Figure 1-2-7.

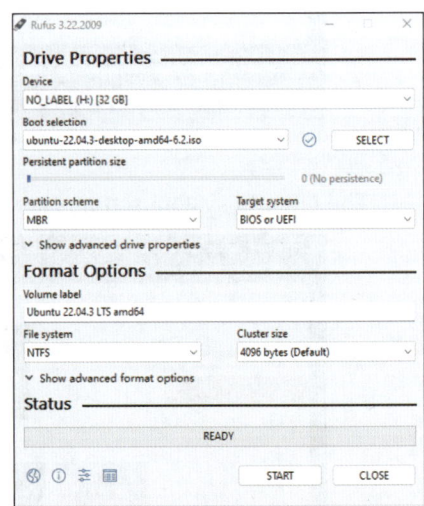

Figure 1-2-6　Rufus settings

Note: the compressed hard disk space should not be allocated. Ubuntu will automatically install to the unallocated hard disk area. If it is allocated, it needs to be reset during installation.

● Video

Ubuntu system installation

Chapter1　Maze Robot Simulation Foundation　**059**

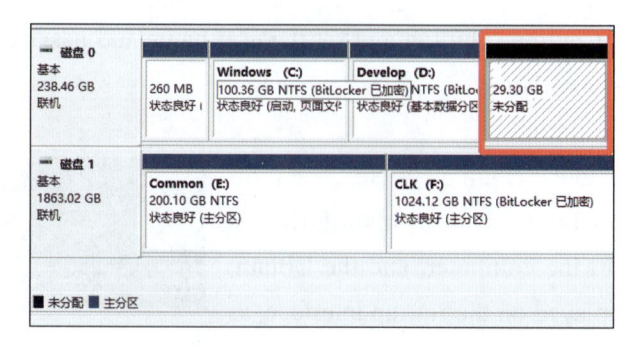

Figure 1-2-7　Compress hard disk

Finally, you need to set the BIOS, and set the first boot method to USB boot.

Different brands of computers have different ways to enter the BIOS, please search by yourself. After setting, save and restart the computer.

2. Install Ubuntu

After restarting the computer, it will enter the USB disk boot mode, select the second "Install Ubuntu", as shown in Figure 1-2-8.

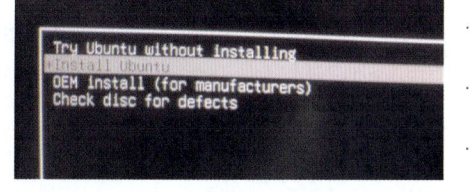

Figure 1-2-8　Install Ubuntu

Select language "Chinese" , "English (US) " for the keyboard, and "Normal installation" for updates and other software.

In the installation type shown in Figure 1-2-9, if you want to install dual systems, select the first "Install Ubuntu, coexisting with Windows Boot Manager", if you want to overwrite Windows, select the second "Clear the entire disk and install Ubuntu".

Figure 1-2-9　Installation type

Click the "Continue" button and set the user name and password at the end to complete the installation.

3. Update Ubuntu

First log in to the Ubuntu system.

After the Ubuntu system installation is complete, restart the computer, and the Ubuntu boot will be displayed on the startup interface, as shown in Figure 1-2-10.

Figure 1-2-10　Ubuntu boot

The first item is to start the Ubuntu system, and the third item is to start the Windows system. Ubuntu is started by default, and the countdown is 10 seconds. You can use the arrow 【 ↑ 】 or 【 ↓ 】 on the keyboard or the mouse to choose to start the Windows system.

First open the software and update, modify the download source to domestic source, such as Huawei source, Tsinghua source, etc. , as shown in Figure 1-2-11.

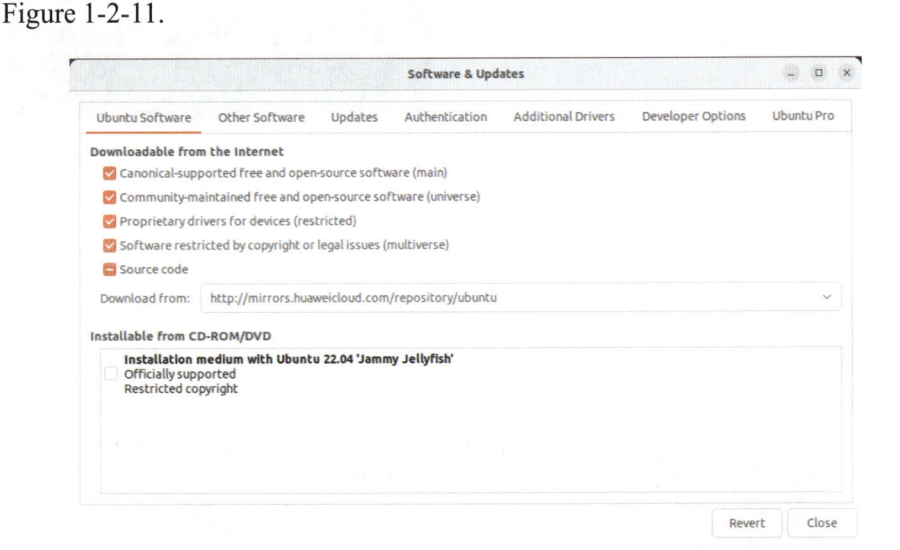

Figure 1-2-11　Modify download source

Open the software updater and update the system. Then set the language.

The Ubuntu system uses English by default, and Chinese can be installed if necessary.

As shown in Figure 1-2-12, click "1" to install the Chinese language, and then set the language to Chinese in "2".

Chapter1　Maze Robot Simulation Foundation　061

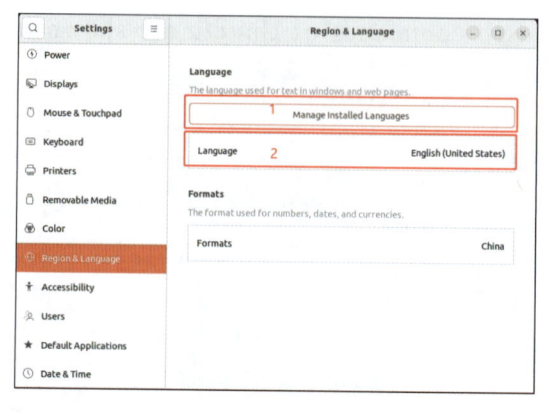

Figure 1-2-12　Install Chinese

Finally install the software you need to use.

Ubuntu has a built-in software store, and you can install software according to your needs. It is recommended to install Visual Studio Code, referred to as VSCode, which is displayed as code in the Ubuntu system software store, as shown in Figure 1-2-13.

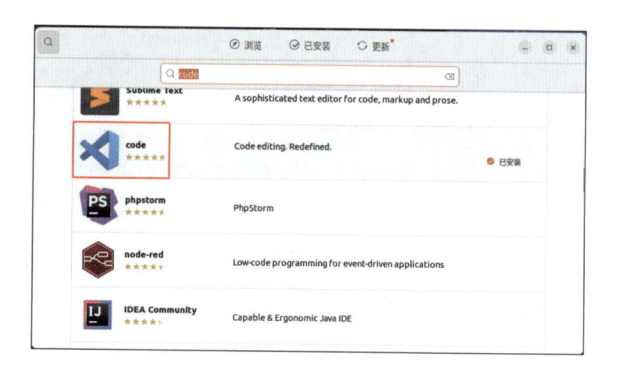

Figure 1-2-13　Install VSCode

Ⅲ. Ubuntu common commands

The Ubuntu system used in this book includes a graphical desktop, which can perform various operations like a Windows system. But it is more commonly used to operate through the terminal. Right-click on any directory and select "Open in Terminal", or use the shortcut key【Ctrl+Alt+T】to start the terminal, as shown in Figure 1-2-14.

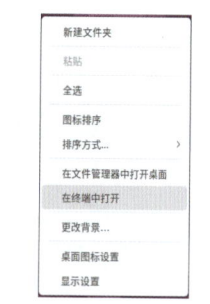

Figure 1-2-14　Right-click menu

Commonly used commands of Ubuntu are shown in Table 1-2-1 to Table 1-2-5.

Table 1-2-1   System commands

Command	Description
shutdown	shutdown
poweroff	shutdown
reboot	reboot

Table 1-2-2   Process commands

Command	Description
ps	Displays the current process
top	Displays the resource usage of each process in the system in real time, similar to Windows task manager
kill	Terminates the specified process

Table 1-2-3   Directory commands

Command	Description
pwd	Shows working directory
cd	Changes shell working directory
ls	Displays the contents of the current working directory
touch	Modifies the access time and modification time of the file. If the file does not exist, it will be automatically created
mkdir	Create directory command
rm	Delete files or directories

Table 1-2-4   User commands

Command	Description
su	User switching command
sudo	Ordinary users temporarily have root permissions
passwd	Changes the password. You must have root permissions to change it

Table 1-2-5   Package management commands

Command	Description
pip	Python-specific commands for installing third-party libraries
apt-get	Used to search, install, upgrade, and uninstall software from the Ubuntu software repository

Chapter1　Maze Robot Simulation Foundation　063

Task 2　Learn ROS2 Installation

The purpose of this task is to have a certain understanding of ROS2 and learn how to install ROS2.

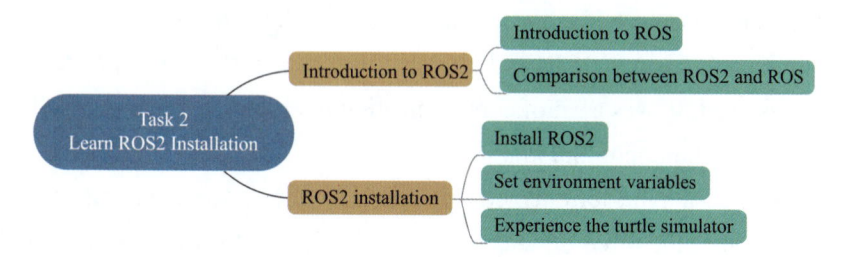

Ⅰ. Introduction to ROS2

1. Introduction to ROS

Born in 2007, ROS is a framework suitable for robot programming. As shown in Figure 1-2-15, this framework couples the original loose components together and provides them with a communication framework. Although ROS is called an operating system, it is not an

::: ROS.org

Figure 1-2-15　ROS

operating system in the usual sense like Windows and Mac. It only connects the operating system and the developed ROS application, so it is also a middleware, and communication is established between ROS-based applications. The bridge, so it is also a runtime environment running on Linux. In this environment, the robot's perception, decision-making, and control algorithms can be better organized and run.

ROS has the following main features:

(1) Distributed, point-to-point

ROS adopts a distributed network framework and uses a communication method based on TCP/IP to realize a point-to-point loosely coupled connection between modules, and can perform various types of communication. The process of the robot can be run separately through the point-to-point design, which is convenient for modular modification and customization, and improves the fault tolerance of the system.

(2) Support multiple languages

ROS supports multiple programming languages. C++ and Python are

currently the most widely used ROS development languages. In addition, ROS also supports many different languages such as LISP, C#, Java, and Octave. In order to support the transplantation and development of more applications, ROS uses a language-neutral interface definition language to implement message transmission between modules. The popular understanding is that the communication format of ROS has nothing to do with which programming language is used. It uses a set of communication interfaces defined by itself.

(3) Rich component tool kits

ROS uses a componentized approach to integrate existing tools and software, such as the three-dimensional visualization platform Rviz in ROS. Rviz is a graphical tool that comes with ROS, which can easily perform graphical operations on ROS programs; Gazebo, a commonly used physics simulation platform, can create a virtual robot simulation environment and add some required parameters to the simulation environment.

(4) Free and open source

ROS has a large open source community ROS WIKI. The application codes in ROS WIKI are classified by maintainers and mainly include core library parts designed and maintained by Willow Garage and some developers, as well as global open source codes developed and maintained by ROS community organizations in different countries. Currently, there are tens of millions of software packages developed using ROS, and there are thousands of related robots. In addition, ROS complies with the BSD license, allowing users to modify and redistribute the application code, and it is completely free for personal and commercial applications and modifications.

2. Comparison between ROS2 and ROS

ROS has been used in a variety of robots since its release, including wheeled robots, legged robots, industrial arms, outdoor unmanned vehicles, self-driving cars, drones, unmanned boats, and more. ROS has evolved from an initial academic research project into a commercial project widely used in industry, and has become a de facto standard in the field of robotics.

With the expansion of application scenarios, the limitations of ROS's initial design have been unable to meet more new needs, such as real-time performance and network latency. ROS2 was born!

ROS2 is a second-generation robot operating system designed and developed based on ROS. It is a software library and tool set that can help simplify robot development tasks and accelerate robot implementation. Compared with ROS, ROS2 has the following three most significant features:

(1) Architecture subversion

Under the ROS architecture, all nodes need to be managed using the Master, which means that before starting the node, it must be registered with the ROS_ Master before it can be used.

ROS2 uses the Discovery mechanism based on DDS. There is no need to use roscore to start the Master, just start the node directly.

(2) API redesign

Most of the code in ROS is based on the API designed in February 2009.

ROS2 has redesigned the user API, but the usage method is similar. ROS2 can use Python3.

(3) Compilation system upgrade

ROS uses rosbuild and catkin to manage projects.

ROS2 uses upgraded versions of ament and colcon, as shown in Table 1-2-6.

Table 1-2-6　ROS2 update plan

Distro	Release date	EOL date
Humble Hawksbill	May 23rd, 2022	May 2027
Galactic Geochelone	May 23rd, 2021	December 9th, 2022
Foxy Fitzroy	June 5th, 2020	May 2023
Eloquent Elusor	November 22nd, 2019	November 2020
Dashing Diademata	May 31st, 2019	May 2021
Crystal Clemmys	December 14th, 2018	December 2019
Bouncy Bolson	July 2nd, 2018	July 2019
Ardent Apalone	December 8th, 2017	December 2018
beta3	September 13th, 2017	December 2017
beta2	July 5th, 2017	September 2017
beta1	December 19th, 2016	July 2017
alpha1 - alpha8	August 31th, 2015	December 2016

II . ROS2 installation

This book uses ROS2 Humble as the preferred version. The installation

process requires a network connection. The installation steps are as follows.

1. Install ROS2

First, replace the domestic source, such as Huawei source, the installation process will be more convenient.

```
$ echo "deb [arch=$(dpkg --print-architecture)] https://repo.
huaweicloud.com/ros2/ubuntu/(lsb_release -cs)main" | sudo tee /
etc/apt/sources.list.d/ros2.list > /dev/null
```

Add security key.

```
$ sudo apt install curl gnupg2 -y
$ curl -s https://gitee.com/clk_china/rosdistro/raw/master/
ros. asc | sudo apt-key add -
```

Update the system.

```
$ sudo apt-get update
$ sudo apt-get upgrade
```

Install ROS2 Humble.

```
$ sudo apt install ros-humble-desktop
```

Install additional applications.

```
$ sudo apt install python3-argcomplete -y
```

2. Set environment variables

After the installation is complete, add the ros2 command to the environment variable and you can directly use ROS2 related commands.

```
$ echo "source /opt/ros/humble/setup. bash" >> ~/.bashrc
```

Compile the configuration file and make it effective immediately.

```
$ source ~/.bashrc
```

Enter ros2.

```
$ ros2
```

The following information is output in the terminal.

```
usage: ros2 [-h] [--use-python-default-buffering]
            Call 'ros2 <command> -h' for more detailed usage.···
```

Chapter1　Maze Robot Simulation Foundation　067

```
ros2 is an extensible command-line tool for ROS 2.
...
```

It means that the ROS2 Humble installation has been completed.

3. Experience the turtle simulator

The TurtleSim is a small game built into ROS2, as shown in Figure 1-2-16. Below we use the command to start it:

```
$ ros2 run turtlesim turtlesim_node
```

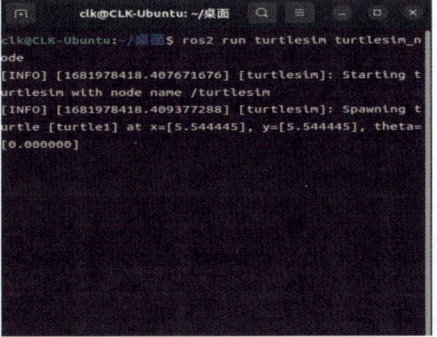 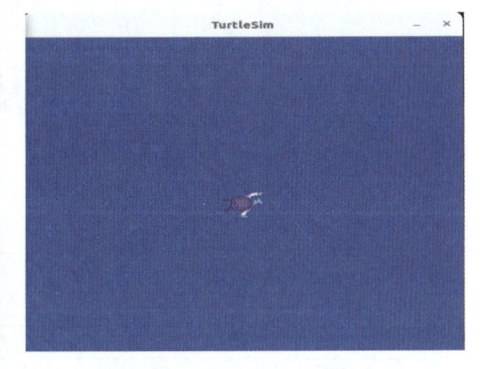

Figure 1-2-16　TurtleSim

Open a new terminal and start the turtle control node, as shown in Figure 1-2-17.

```
$ ros2 run turtlesim turtle_teleop_key
```

Figure 1-2-17　Turtle control node

Use the up, down, left and right arrows to move the turtle, use 【 E 】【 R 】【 T 】【 D 】【 G 】【 C 】【 V 】【 B 】 to rotate the turtle, use 【 F 】 to stop the rotation, and use 【 Q 】 to exit.

Task 3　Learn Gazebo Installation and Integration

The purpose of this task is to have a certain understanding of Gazebo and learn how to install Gazebo.

Video ●

Experience the turtle simulator

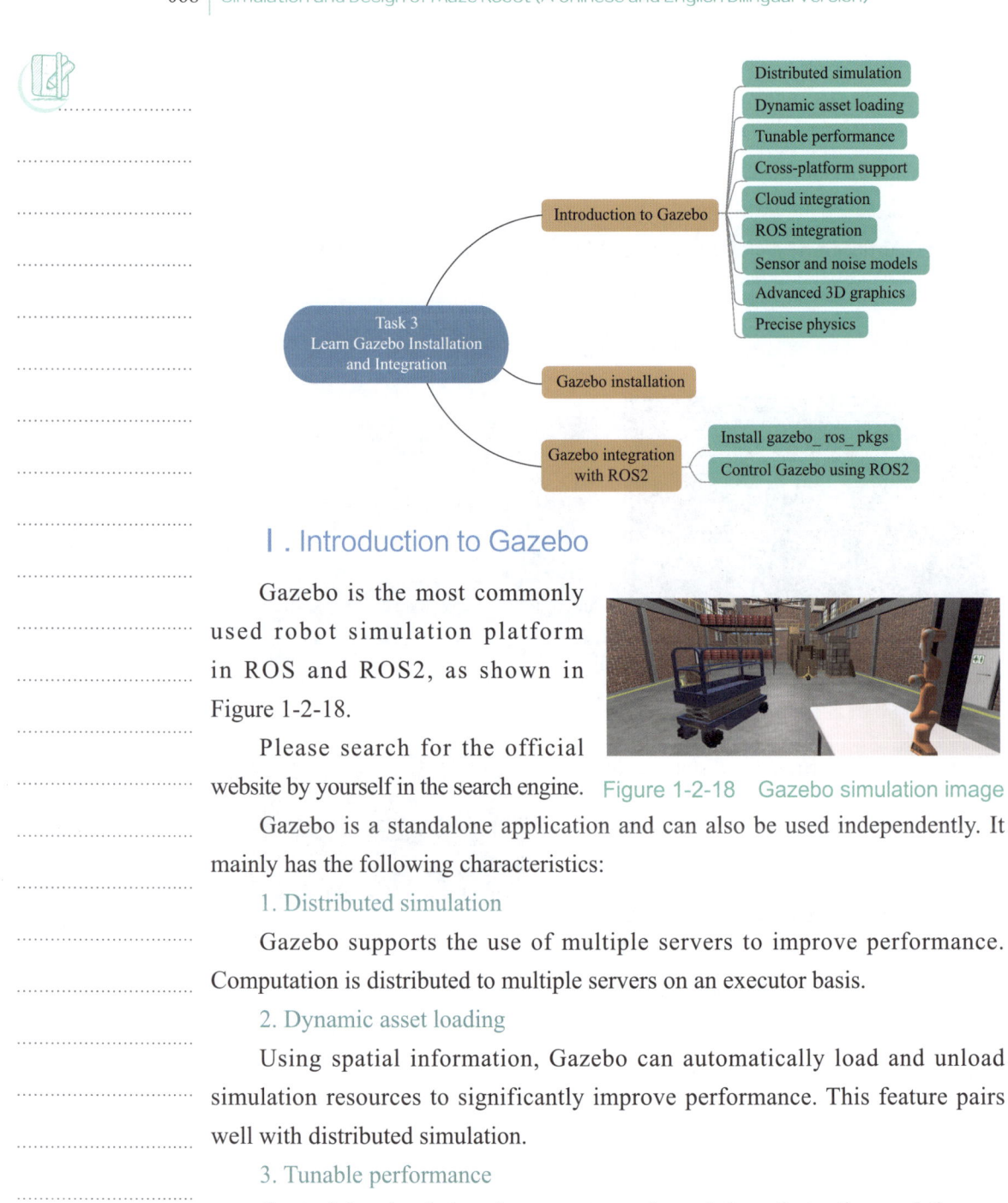

Ⅰ. Introduction to Gazebo

Gazebo is the most commonly used robot simulation platform in ROS and ROS2, as shown in Figure 1-2-18.

Please search for the official website by yourself in the search engine.

Figure 1-2-18　Gazebo simulation image

Gazebo is a standalone application and can also be used independently. It mainly has the following characteristics:

1. Distributed simulation

Gazebo supports the use of multiple servers to improve performance. Computation is distributed to multiple servers on an executor basis.

2. Dynamic asset loading

Using spatial information, Gazebo can automatically load and unload simulation resources to significantly improve performance. This feature pairs well with distributed simulation.

3. Tunable performance

Control the simulation time step to run in real time, faster than real time, or slower than real time.

4. Cross-platform support

Use the Gazebo library on Windows, Linux and MacOS.

5. Cloud integration

View, download and upload simulation models and scene model worlds on our cloud-hosted server app. gazebosim. org.

6. ROS integration

Supports communication with ROS Melodic and allows for convenient data conversion

7. Sensor and noise models

Monocular cameras, depth cameras, lidar, inertial sensor(IMU), contact sensors, altimeter sensors and magnetometer sensors are all available, with more in development. Each sensor can optionally utilize noise models to inject Gaussian or customized noise properties.

8. Advanced 3D graphics

Gazebo supports OGRE 2.1 rendering.OGRE 2.1 is the rendering engine, version 2.1, which provides access to the latest rendering technologies, including enhanced shadow maps, PBR materials and faster rendering pipelines.

9. Precise physics

DART is the default physics engine in Gazebo physics, providing accuracy beyond game engines.

Ⅱ. Gazebo installation

This book uses the latest Gazebo 11 as the simulation environment. The installation command is as follows:

```
$ sudo apt install gazebo
```

If you want to run Gazebo, enter gazebo directly in the terminal, as shown in Figure 1-2-19.

```
$ gazebo
```

Gazebo officially provides a large number of models for use. Use the following command to install them.

```
$ cd ~/.gazebo && wget https://gitee.com/clk_china/scripts/
raw/master/gazebo_model. py && python3 gazebo_model.py
```

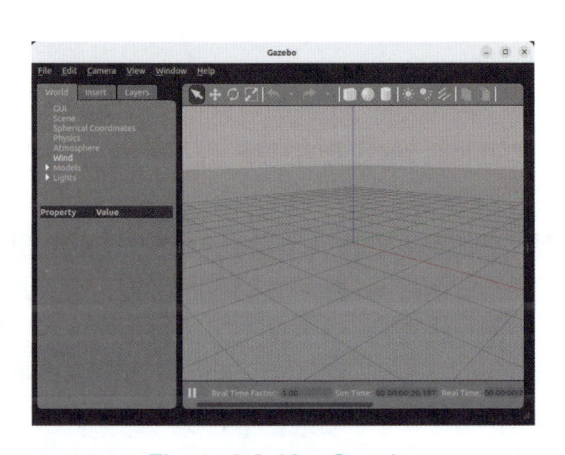

Figure 1-2-19　Gazebo

III. Gazebo integration with ROS2

Gazebo is an independent software. To integrate with ROS2, some dependencies need to be installed.

1. Install gazebo_ros_pkgs

gazebo_ros_pkgs is the bridge between ROS2 and Gazebo, as shown in Figure 1-2-20. Use the following command to install it with one click.

```
$ sudo apt install ros-humble-gazebo-*
```

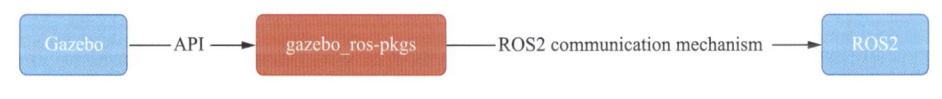

Figure 1-2-20　Integration principle of Gazebo and ROS2

2. Control Gazebo using ROS2

First start Gazebo and load a world file, as shown in Figure 1-2-21.

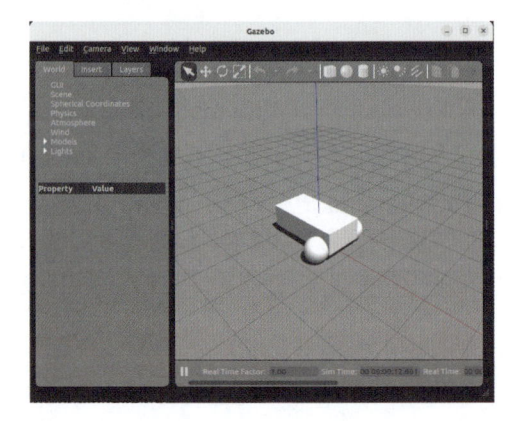

Figure 1-2-21　Gazebo starts the specified world file

Chapter1 Maze Robot Simulation Foundation | 071

```
$ gazebo /opt/ros/humble/share/gazebo_plugins/worlds/gazebo_
ros_diff_drive_demo.world
```

Gazebo will directly load the specified world file when starting and display a two-wheeled car.

Use the ROS2 command to drive the car and open a new terminal. The command is as follows:

```
$ ros2 topic pub /demo/cmd_demo geometry_msgs/msg/Twist
"{linear: {x: 0.2,y: 0,z: 0},angular: {x: 0,y: 0,z: 0}}"
```

The following information is output in the terminal:

```
publisher: beginning loop
publishing #1: geometry_msgs. msg.Twist(linear=geometry_msgs.
msg.Vector3(x=0.2, y=0.0, z=0.0),
  angular=geometry_msgs.msg.Vector3(x=0.0, y=0.0, z=0.0))

publishing #2: geometry_msgs. msg. Twist(linear=geometry_msgs.
msg. Vector3(x=0.2, y=0.0, z=0.0), angular=geometry_msgs. msg.
Vector3(x=0.0, y=0.0, z=0.0))

publishing #3: geometry_msgs.msg.Twist(linear=geometry_msgs.
msg.Vector3(x=0.2, y=0.0, z=0.0), angular=geometry_msgs.msg.
Vector3(x=0.0, y=0.0, z=0.0))
  ...
```

Check that the car in Gazebo starts to run in a straight line.

Questions and Summary

1. What are the commonly used commands in Ubuntu system?

2. How does ROS2 work?

3. ROS2 can be integrated with Gazebo through special tools, which will be very convenient for model simulation and control.

Video

Simulation
environment
integration

Project 3

Getting Started with ROS2

Learning objectives

1. Master how to create ROS2 workspace and function packages.
2. Master how ROS2 works.
3. Master the basic communication methods of ROS2.

Task 1   Understand the ROS2 Workspace and Common Commands

The purpose of this task is to have a certain understanding of the working mode of ROS2 and learn to use the ROS2 workspace and function packages.

I. Workspace

When using some development environments, such as Visual Studio, Eclipse, Qt Creator, Keil, etc., a project will be created before starting to write a program. At this time, a folder will be generated, and all files for this project will be placed in this folder. This folder and the contents inside are called projects, as shown in Figure 1-3-1.

When developing in ROS2, various code files, parameter configuration files, etc. also need to be placed in a folder for management. This folder is a function

Chapter1　Maze Robot Simulation Foundation | 073

package. The workspace is a directory containing several function packages, and ROS2 relies on the workspace to find function packages.

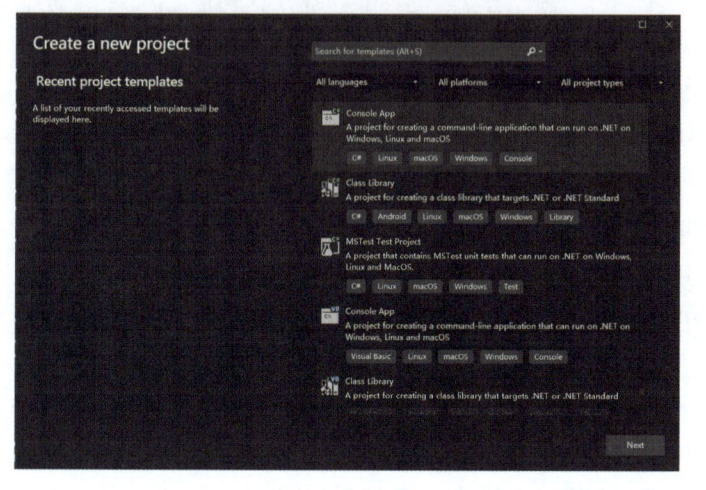

Figure 1-3-1　Creating a project with Visual Studio

1. Create a workspace

The workspace must contain a src folder to store the ROS2 function package. Create a workspace named ros2_ws with the following command:

```
$ mkdir -p ~/ros2_ws/src
```

2. Create packages

A package is a directory containing certain specific functions that ROS2 can execute directly.

The ros2 pkg command is used to create packages.

```
$ ros2 pkg create <package> --build-type <type> --dependencies
<dependency> --node-name <nodename>
```

The parameters in the command are shown in Table 1-3-1.

Table 1-3-1　ros2 pkg commands

Command	Describe
create	Create packages. The <package> naming convention is to avoid capital letters
--build-type	Compiled type, cmake|ament_cmake|ament_python Python select ament_python
--dependencies	Dependencies, python at least add rclpy

Continued

Command	Describe
--node-name	For the added node name, the node file will be automatically created and the node will be declared in setup.py. Only one node can be set. If more than one node is used, it needs to be added manually

Function packages in ROS2 are divided into three types according to different compilation methods:

① ament_python: suitable for Python programs;

② cmake: suitable for C++ programs;

③ ament_cmake: suitable to C++ programs, it is an enhanced version of cmake.

Function packages in ROS2 mainly consist of two dependencies depending on the language type:

① rclpy: suitable for Python dependencies;

② rclcpp: suitable for C++ dependencies.

For example, create a function package named pythondemo.

```
$ cd ~/ros2_ws/src
$ ros2 pkg create pythondemo --build-type ament_python --dependencies
rclpy
```

The pythondemo directory structure after creation:

```
pythondemo
├── pythondemo
│   └── __init__.py
├── package.xml
├── resource
│   └── pythondemo
├── setup.cfg
├── setup.py
└── test
    ├── test_copyright.py
    ├── test_flake8.py
    └── test_pep257.py

3 directories, 8 files
```

3. Build packages

Whenever there are modifications to the function package, it needs to be rebuilt so that the workspace can recognize the function package.

Chapter1　Maze Robot Simulation Foundation | 075

ROS2 uses colcon as a function package building tool, which is equivalent to the catkin tool in ROS. ROS2 does not have colcon installed by default, so it needs to be installed first:

```
$ sudo apt-get install python3-colcon-common-extensions
```

Compile the workspace:

```
$ cd ~/ros2_ws
$ colcon build
```

All packages under src will be automatically built at once:

```
Starting >>> pythondemo
--- stderr: pythondemo
/usr/lib/python3/dist-packages/setuptools/command/install.py:
34: SetuptoolsDeprecationWarning: setup.py install is deprecated.
Use build and pip and other standards-based tools.
    warnings.warn(
---
Finished <<< pythondemo [0.67s]

Summary: 1 package finished [0.82s]
    1 package had stderr output: pythondemo
```

It doesn't matter if you see the above program error in your build, it will not affect use, ROS2 is officially fixing it. The cause of the error is that the setuptools version is too high.

After the build is completed, build, install and log directories will be added to the directory at the same level as src:

```
.
├── build
├── install
├── log
└── src

4 directories, 0 files
```

① The build directory stores intermediate files. For each package, a subfolder is created in which e. g. cmake is called.

② The install directory is where each package will be installed. By default, each package will be installed into a separate subdirectory.

③ The log directory contains various log information about each colcon call.

So far, we have only created and built the function package. We will learn to add nodes and publish and subscribe topics in Task 2 and Task 3 of this project.

II. Common commands

When creating the function package, we used the ros2 pkg command. Some other commonly used commands are shown in Table 1-3-2 to Table 1-3-5. You can add --help to view detailed usage, such as ros2 pkg create --help.

Table 1-3-2   Package commands

Command	Description
ros2 pkg create	Create a function package
ros2 pkg list	List all function packages
ros2 pkg executables	List the executable files of the function package

Table 1-3-3   Node commands

Command	Describe
ros2 node list	View the list of currently active nodes
ros2 node info	View node details, including subscriptions, published messages, enabled services and actions, etc.
ros2 run	Node running tools
ros2 launch	launch tool, which can start multiple nodes at one time

Table 1-3-4   Topic commands

Command	Describe
ros2 topic list	Returns a list of all topics currently active in the system
ros2 topic type	View topic message type
ros2 topic info	View topic information, including message type, number of subscribers, number of publishers, etc.
ros2 topic echo	Display topic content on the console in real time
ros2 topic pub	Manually post to a topic

Table 1-3-5   Interface commands

Command	Describe
ros2 interface list	Display all interfaces in the system, including Messages, Services, and Actions
ros2 interface show	Display details of the specified interface

Chapter1 Maze Robot Simulation Foundation | 077

Task 2　Learn ROS2 Nodes

The purpose of this task is to have a certain concept of ROS2 nodes and learn how to create and use ROS2 nodes.

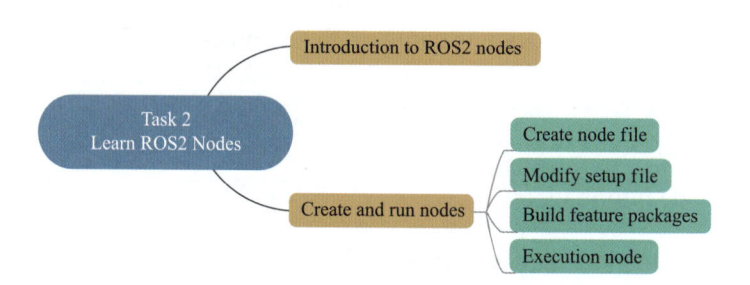

I . Introduction to ROS2 nodes

ROS2 nodes are executable files used to implement specific functions in the ROS2 function package. A ROS2 function package can contain multiple ROS2 nodes. In this book, the function packages are all written in Python, so the node is a py file.

A node is the smallest unit in ROS2. A node can publish or receive a topic, and a node can also provide or use a certain service.

Open two terminals and start the TurtleSim and control node respectively, as shown in Figure 1-3-2.

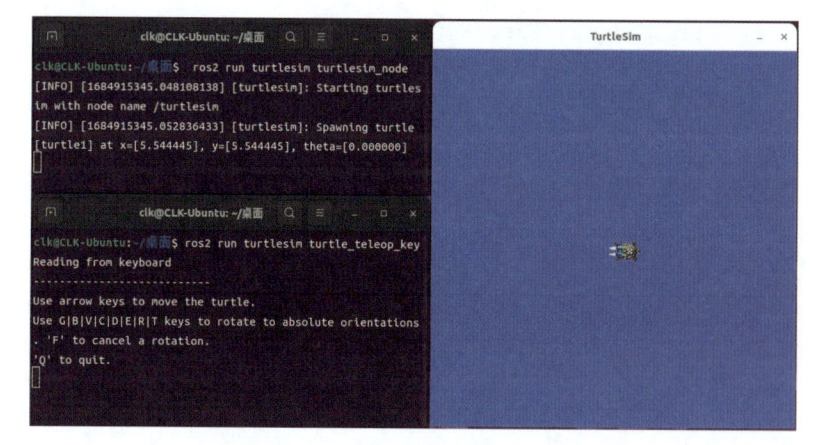

Figure 1-3-2　TurtleSim and control nodes

The ros2 run command is used to start the node.

```
$ ros2 run turtlesim turtlesim_node
```

```
$ ros2 run turtlesim turtle_teleop_key
```

View the list of currently active nodes:

```
$ ros2 node list
```

The following information is output in the terminal:

```
/teleop_turtle
/turtlesim
```

View node information:

```
$ ros2 node info /turtlesim
```

The following information is output in the terminal:

```
/turtlesim
  Subscribers:
    /parameter_events: rcl_interfaces/msg/ParameterEvent
    /turtle1/cmd_vel: geometry_msgs/msg/Twist
  Publishers:
    /parameter_events: rcl_interfaces/msg/ParameterEvent
    /rosout: rcl_interfaces/msg/Log
    /turtle1/color_sensor: turtlesim/msg/Color
    /turtle1/pose: turtlesim/msg/Pose
  Service Servers:
    /clear: std_srvs/srv/Empty
    /kill: turtlesim/srv/Kill
    /reset: std_srvs/srv/Empty
    /spawn: turtlesim/srv/Spawn
    /turtle1/set_pen: turtlesim/srv/SetPen
    /turtle1/teleport_absolute: turtlesim/srv/TeleportAbsolute
    /turtle1/teleport_relative: turtlesim/srv/TeleportRelative
    /turtlesim/describe_parameters: rcl_interfaces/srv/Describe
Parameters
        /turtlesim/get_parameter_types: rcl_interfaces/srv/GetParame
terTypes
    /turtlesim/get_parameters: rcl_interfaces/srv/GetParameters
    /turtlesim/list_parameters: rcl_interfaces/srv/ListParameters
    /turtlesim/set_parameters: rcl_interfaces/srv/SetParameters
     /turtlesim/set_parameters_atomically: rcl_interfaces/srv/
SetParametersAtomically
  Service Clients:

  Action Servers:
    /turtle1/rotate_absolute: turtlesim/action/RotateAbsolute
  Action Clients:
```

Ⅱ. Create and run nodes

Next, we will create our own node in the function package created in Task 1 of this project.

The node file must be placed in the same directory as __init__. py.

1. Create node file

In the same directory as __init__. py, create a file named demo. py. The file name can be customized, but please be careful not to include Chinese and Chinese characters.

```python
#!/usr/bin/env python3
import rclpy
from rclpy. node import Node

class Demonode(Node):
    def __init__(self,name):
        super().__init__(name)
        self.count = 0
        self.get_logger().info("{}node started". format(name) )
        #create a timer callback with a timing interval of 0.5s
        self. timer = self. create_timer(0.5, self.timer_callback)

    def timer_callback(self):
        """timer callback function"""
        self.get_logger().info('this is the {}th callback'.format
(self.count)) #Print the published data
        self. count += 1


def main(args=None):
    rclpy. init()              # initialize rclpy
    node = Demonode("demo")   # create a new node
    rclpy. spin(node)         # keep the node running and check
                              # whether the exit command is
                              # received (Ctrl+C)
    rclpy. shutdown()         # close rclpy
```

2. Modify setup file

After creating the node, you need to add a statement in setup. py so that the node can be found during build.

As shown below, add 'demo=pythondemo. demo:main', which means that an executable file demo will be generated after building, pointing to the main function in pythondemo/demo. py.

```
entry_points={
    'console_scripts': [
'demo = pythondemo. demo:main'
    ],
},
```

3. Build feature packages

After modifying the node, you need to rebuild the function package.

```
$ cd ~/ros2_ws
$ colcon build
```

4. Execution node

Before executing the function package, you need to source the environment variables in the current workspace. Environment variables are only valid for the current terminal. If you open a new terminal, you need to resource it.

```
$ source install/setup.bash
$ ros2 run pythondemo demo
```

The following information is output in the terminal:

```
[INFO] [1681981630. 122434895] [demo]: The demo node is started
[INFO] [1681981630. 623920341] [demo]: This is the 0th callback
[INFO] [1681981631. 123787430] [demo]: This is the 1st callback
[INFO] [1681981631. 624489612] [demo]: This is the 2nd callback
[INFO] [1681981632. 123899354] [demo]: This is the 3rd callback
[INFO] [1681981632. 623877488] [demo]: This is the 4th callback
[INFO] [1681981633. 123852965] [demo]: This is the 5th callback
[INFO] [1681981633. 624441352] [demo]: This is the 6th callback
```

Print the node list:

```
$ ros2 node list
```

The following information is output in the terminal:

```
/demo
```

Task 3   Learn ROS2 Topic

The purpose of this task is to have a certain understanding of ROS2 topics and learn how to publish and subscribe to messages through topics.

Chapter1　Maze Robot Simulation Foundation　081

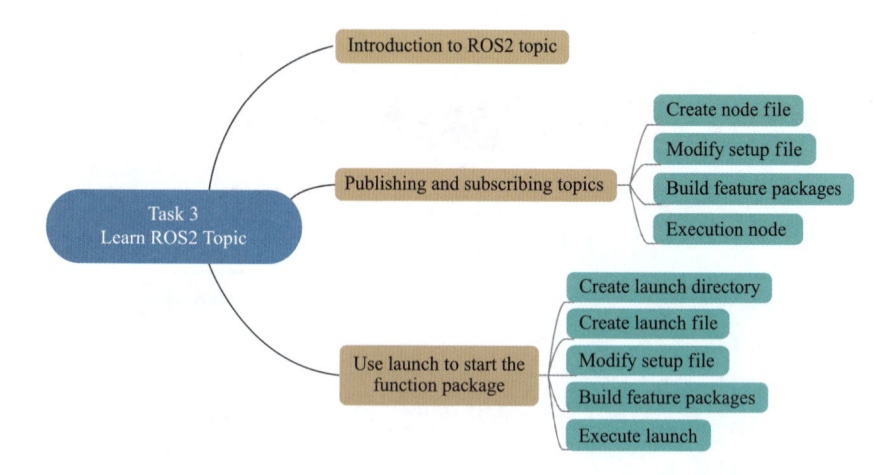

Ⅰ. Introduction to ROS2 topic

Topics are the most commonly used communication method in ROS2. As shown in Figure 1-3-3, lidar, differential drive and camera all transmit data through topics.

When one node publishes data to a topic, another node can get the data by subscribing to the topic.

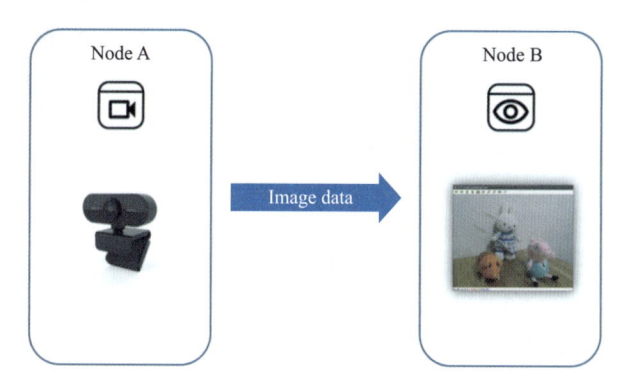

Figure 1-3-3　Topic communication

ROS2 topic communication supports multiple modes, which can be 1 to 1, as shown in Figure 1-3-4; 1 to n, as shown in Figure 1-3-5; n to n, as shown in Figure 1-3-6; n to 1, as shown in Figure 1-3-7; or even subscribe to topics published by itself, as shown in Figure 1-3-8.

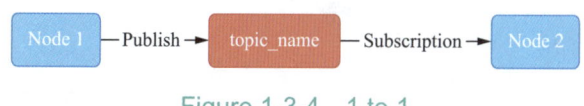

Figure 1-3-4　1 to 1

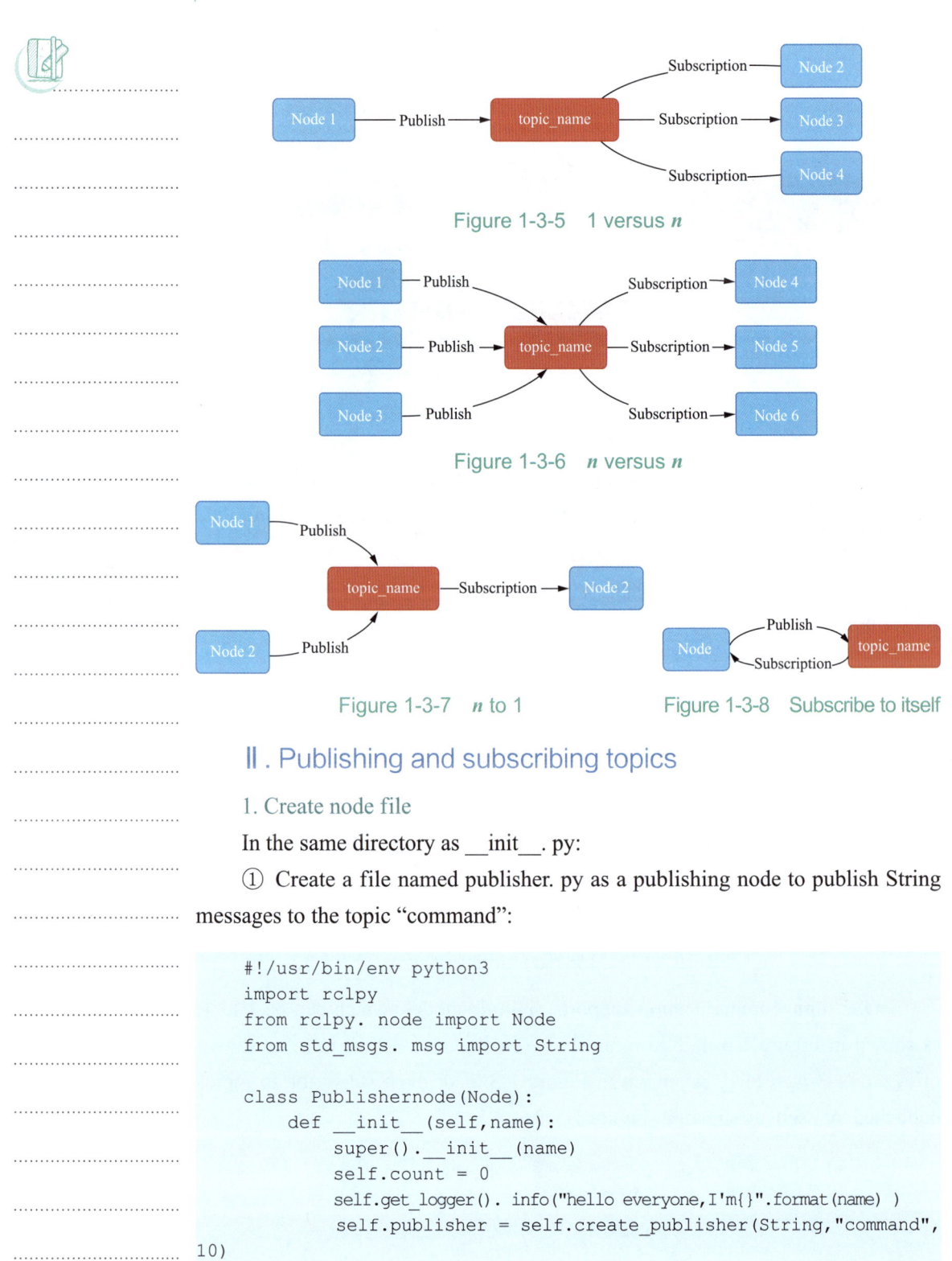

Figure 1-3-5   1 versus *n*

Figure 1-3-6   *n* versus *n*

Figure 1-3-7   *n* to 1

Figure 1-3-8   Subscribe to itself

II. Publishing and subscribing topics

1. Create node file

In the same directory as __init__. py:

① Create a file named publisher. py as a publishing node to publish String messages to the topic "command":

```
#!/usr/bin/env python3
import rclpy
from rclpy. node import Node
from std_msgs. msg import String

class Publishernode(Node):
    def __init__(self,name):
        super().__init__(name)
        self.count = 0
        self.get_logger(). info("hello everyone,I'm{}".format(name) )
        self.publisher = self.create_publisher(String,"command",
10)
```

Chapter1　Maze Robot Simulation Foundation | 083

```python
        #create a timing callback with a timing interval of 0.5 s
        self. timer = self.create_timer(0.5, self.timer_callback)

    def timer_callback(self):
        """ timer callback function """
        msg = String()
        msg.data = 'hello {}'.format(self.count)
        self.publisher.publish(msg)        # publish data
        self.get_logger(). info(f'publish command: {msg.data}')
                                          # print the published data
        self.count += 1


def main(args=None):
    rclpy.init(args=args)                      # initialize rclpy
    node = Publishernode("publisher_node") # create a new node
    rclpy.spin(node)    # keep the node running and check whether
                        # the exit command is received (Ctrl+C)
    rclpy.shutdown()                           # close rclpy
```

② Create a file named subscriber. py as a subscription node, subscribe to the topic "command", and the message type is String:

```python
#!/usr/bin/env python3
import rclpy
from rclpy.node import Node
from std_msgs.msg import String

class Subscribernode(Node):
    def __init__(self,name):
        super(). __init__(name)
        self.get_logger().info("hello everyone,I'm{}".format(name))
        # subscribe to topics
        self.subscribe = self. create_subscription(String,"com
mand",self.command_callback,10)

    def command_callback(self, msg):
        self.get_logger().info('received a message,the content
is{}'.format(msg. data))


def main(args=None):
    rclpy.init(args=args)      #initialize rclpy
    node = Subscribernode("subscriber_node")   # create a new node
    rclpy.spin(node)     # keep the node running and check whether
                        # the exit command is received (Ctrl+C)
    rclpy.shutdown()     # close rclpy
```

2. Modify setup file

Add declarations in setup. py to enable ROS2 to find publishing nodes and subscribing nodes.

```
entry_points={
    'console_scripts': [
'publisher = pythondemo.publisher:main',
'subscriber = pythondemo.subscriber:main',
    ],
},
```

Generate an executable file publisher, pointing to the main function in pythondemo/publisher. py.

Generate an executable file subscriber, pointing to the main function in pythondemo/subscriber. py.

3. Build feature packages

After modifying the node, you need to rebuild the function package.

```
$ cd ~/ros2_ws
$ colcon build
```

4. Execution node

Use the source command to load the environment variables in the current workspace and execute the subscription node.

```
$ source install/setup.bash
$ ros2 run pythondemo subscriber
```

The following information is output in the terminal:

```
[INFO] [1682039177.737676796] [subscriber_node]: hello everyone,
I'm subscriber_node
```

No new content will be output. This is because the publishing node has not been started yet and the subscribing node has not obtained the data.

Open a new terminal and execute the publish node, as shown in Figure 1-3-9.

```
$ cd ~/ros2_ws
$ source install/setup.bash
$ ros2 run pythondemo publisher
```

Print the node list.

Chapter1 Maze Robot Simulation Foundation | 085

```
$ ros2 node list
```

Figure 1-3-9 Topic communication

The following information is output in the terminal:

```
/publisher_node
/subscriber_node
```

Print the topic list.

```
$ ros2 topic list
```

The following information is output in the terminal:

```
/command
/parameter_events
/rosout
```

The subscribing node and the publishing node start to update synchronously, thus realizing topic communication.

Ⅲ. Use launch to start the function package

In "Ⅱ. Publishing and subscribing topics", two terminals are used to start the publishing node and the subscription node respectively to realize topic communication. This method of operation is very inconvenient if a project contains a large number of nodes. ROS2 provides another way to start nodes, the ros2 launch command, which can start multiple nodes at one time.

1. Create launch directory

The ros2 launch command uses launch files to launch multiple nodes at once. ROS2 officially recommends placing launch files in the launch folder for easy maintenance.

Enter the function package directory and create the launch folder.

```
$ cd ~/ros2_ws/src/pythondemo
```

```
$ mkdir -p launch
```

2. Create launch file

The launch file is actually a py file, and the officially recommended naming method is ***. launch. py.

Use VScode to create a file named pythondemo.launch.py in launch and write the following code:

```
from launch import LaunchDescription
from launch_ros.actions import Node

def generate_launch_description():
    """
    1.preserved name, ROS2 will recognize these function names
    2.launch content description function, called by ros2 launch
through scanning
    """
    publisher = Node(
        package="pythondemo",
        executable="publisher"
        )
    subscriber = Node(
        package="pythondemo",
        executable="subscriber"
        )
    # create a LaunchDescription object launch_description to describe
    # the launch file
    launch_description = LaunchDescription([publisher, subscriber])
    # return to let ROS2 execute the node according to the launch
    # description
    return launch_description
```

3. Modify setup file

The launch file is a data file relative to the function package, and its installation must be added to the setup file, otherwise the launch file will not be found after building. The complete modified setup file is as follows:

```
from setuptools import setup
from glob import glob
import os
package_name = 'pythondemo'

setup(
    name=package_name,
    version='0.0.0',
```

```python
        packages=[package_name],
        data_files=[
            ('share/ament_index/resource_index/packages',
                ['resource/' + package_name]),
            ('share/' + package_name, ['package. xml']),
            (os.path.join('share', package_name, 'launch'), glob
('launch/*py') ) ,          # add installation of launch files
        ],
        install_requires=['setuptools'],
        zip_safe=True,
        maintainer='clk',
        maintainer_email='clk@todo.todo',
        description='TODO: Package description',
        license='TODO: License declaration',
        tests_require=['pytest'],
        entry_points={
            'console_scripts': [
        'publisher = pythondemo.publisher:main',
        'subscriber = pythondemo.subscriber:main',
            ],
        },
    )
```

4. Build feature packages

Rebuild the feature package.

```
$ cd ~/ros2_ws
$ colcon build
```

5. Execute launch

Execute launch to load environment variables, and use ros2 launch to start the launch file.

For detailed usage of the ros2 launch command, please view it through ros2 launch --help.

```
$ source install/setup.bash
$ ros2 launch pythondemo pythondemo.launch.py
```

The following information is output in the terminal:

```
[INFO] [launch]: All log files can be found below /home/clk/.
ros/log/2023-04-21-09-30-19-670423-CLK-Ubuntu-6953
[INFO] [launch]: Default logging verbosity is set to INFO
[INFO] [publisher-1]: process started with pid [6954]
[INFO] [subscriber-2]: process started with pid [6956]
[subscriber-2] [INFO] [1682040619. 879770345] [subscriber_
```

```
node]: hello everyone,I'm subscriber_node
    [publisher-1] [INFO] [1682040619. 879963830] [publisher_node]:
hello everyone,I'm publisher_node
    [publisher-1] [INFO] [1682040620. 382284307] [publisher_node]:
publish command : hello 0
    [subscriber-2] [INFO] [1682040620. 382986365] [subscriber_
node]: received a message, the content is hello 0
    [publisher-1] [INFO] [1682040620. 882385035] [publisher_node]:
publish command : hello 1
    [subscriber-2] [INFO] [1682040620. 883157585] [subscriber_
node]: received a message, the content is hello 1
    [publisher-1] [INFO] [1682040621. 381542844] [publisher_node]:
publish command : hello 2
    [subscriber-2] [INFO] [1682040621. 382035717] [subscriber_
node]: received a message, the content is hello 2
    [publisher-1] [INFO] [1682040621. 881919999] [publisher_node]:
publish command : hello 3
    [subscriber-2] [INFO] [1682040621. 883140963] [subscriber_
node]: received a message, the content is hello 3
    [publisher-1] [INFO] [1682040622. 382903308] [publisher_node]:
publish command : hello 4
    [subscriber-2] [INFO] [1682040622. 383772402] [subscriber_
node]: received a message, the content is hello 4
    [publisher-1] [INFO] [1682040622. 882459446] [publisher_node]:
publish command : hello 5
    [subscriber-2] [INFO] [1682040622. 883335239] [subscriber_
node]: received a message, the content is hello 5
    …
```

It can be seen that two nodes are started at one time, realizing topic communication.

Print the node list.

```
$ ros2 node list
```

The following information is output in the terminal:

```
/publisher_node
/subscriber_node
```

Print the topic list.

```
$ ros2 topic list
```

The following information is output in the terminal:

```
/command
/parameter_events
/rosout
```

Chapter1　Maze Robot Simulation Foundation | 089

Project 4

Getting Started with Gazebo

Learning objectives

1. Master the basic operation methods of Gazebo.
2. Master to build 3D models.

Task 1　Master the Basic Operations of Gazebo

The purpose of this task is to have a certain understanding of Gazebo and master the basic operation methods.

Ⅰ. Introduction to Gazebo functions

1. Introduction and use of Gazebo interface

This book has already introduced Gazebo in Project 2 Task 3, now start it directly:

```
$ gazebo
```

As shown in Figure 1-4-1, 1 is the menu bar, 2 is the toolbar, 3 is the left panel, and 4 is the simulation view. The red line, green line and blue line are the coordinate axes of Gazebo, respectively. Corresponds to x-axis, y-axis and z-axis.

(1) Menu bar

Mainly use the File menu and Edit menu.

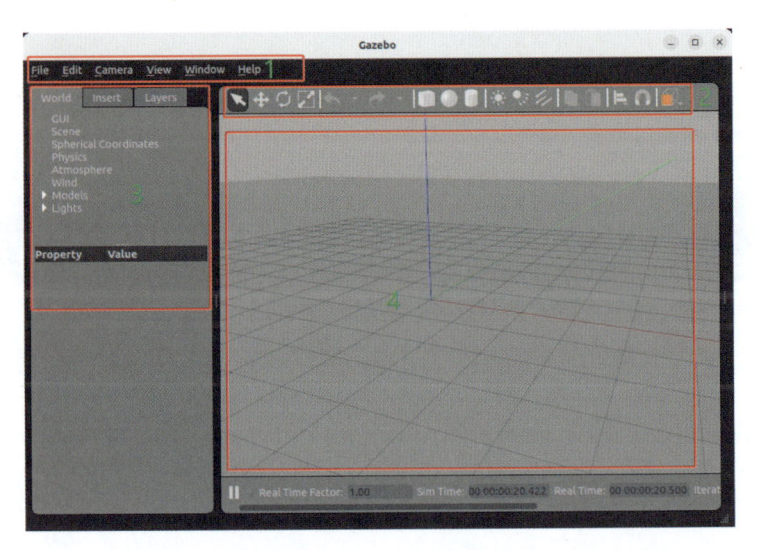

Figure 1-4-1　Gazebo

File menu: Includes save and exit functions, as shown in Figure 1-4-2.

Edit menu: Edit menu, as shown in Figure 1-4-3.

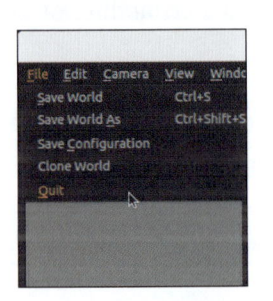

Figure 1-4-2　File menu

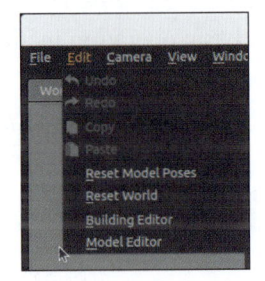

Figure 1-4-3　Edit menu

(2) Toolbar

Contains some functions needed to interact with Gazebo, from left to right, as shown in Table 1-4-1.

Table 1-4-1　Toolbar comparison table

Number	Describe
1	Selection mode: Default mode, mark in the scene
2	Movement mode: You can move the model along the x-axis, y-axis, z-axis, or in any direction
3	Rotation mode: You can rotate the model along the x-axis, y-axis, and z-axis
4	Zoom mode: You can reduce or enlarge the model along the x-axis, y-axis, and z-axis
5	Undo
6	Redo

Chapter1　Maze Robot Simulation Foundation | 091

Continued

Number	Describe
7	Place a rectangle
8	Place a sphere
9	Place a cylinder
10	Place a point light (spherical point light)
11	Place the spotlight (shine from above, in a pyramid shape)
12	Place directional light sources (directional lights)
13	Copy
14	Paste
15	Align: Align models to each other
16	Snap: Snap one model to another model
17	Change View: See the scene from a different angle

(3) Left panel

World: Displays all models in the current scene. You can view and modify their parameters through this option.

Insert: Add a new model to the current scene.

Layers: Organizes and displays the different visualization groups available in the simulation, rarely used.

(4) Simulation view

This is Gazebo's simulation window, where all models are displayed.

2. Gazebo system commands

Execute the Gazebo help command to view the help information.

```
$ gazebo -h
```

Gazebo use method: gazebo [options] <world_file>.

When Gazebo is started, you can specify the world file, which is the two-wheeled car displayed in Project 2 Task 3. At the same time, Gazebo can also add some configuration information when starting. Commonly used configurations are shown in Table 1-4-2.

Table 1-4-2　Commonly used configurations

Parameter	Describe
-v [--version]	Output version information
--verbose	Add messages written to the terminal

Continued	
Parameter	Describe
-h [--help]	Show help information
-u [--pause]	Start the server in a paused state
-e [--physics] arg	Specify a physics engine, ode|bullet|dart|simbody
--gui-client-plugin arg	Load the GUI plugin
-s [--server-plugin] arg	Load server plugin

When starting the ROS2 plug-in, you need to use the -s parameter to add and start the ROS2 plug-in. The command is as follows:

```
$ gazebo --verbose -s libgazebo_ros_init. so -s libgazebo_ros_
factory. so
```

libgazebo_ros_init.so and libgazebo_ros_factory.so are ROS2 plug-ins. You can use ROS2 commands to operate Gazebo after starting them.

Ⅱ. Gazebo drawing operation

1. Drawing models

(1) Add model

Use the left mouse button to click the icon on the toolbar, and then left-click at a suitable location in the simulation view to place the model, as shown in Figure 1-4-4.

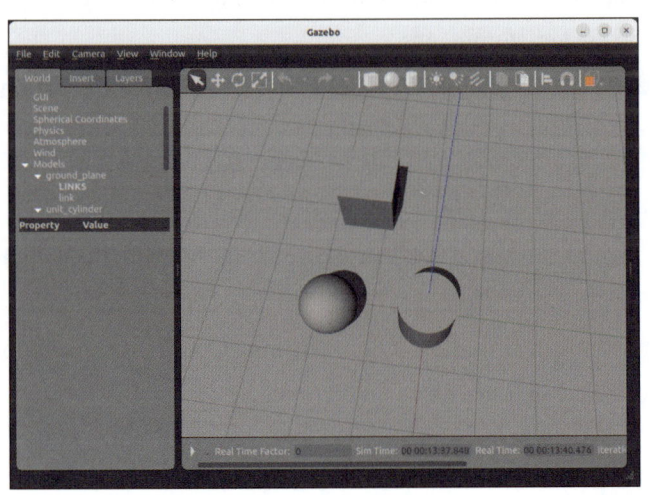

Figure 1-4-4　Add model

Click Move, Rotate or Scale button in the toolbar, as shown in Figure 1-4-5, to

● Video

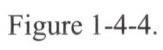

Simulation model drawing

enter the corresponding model and operate the position, direction and size of the model.

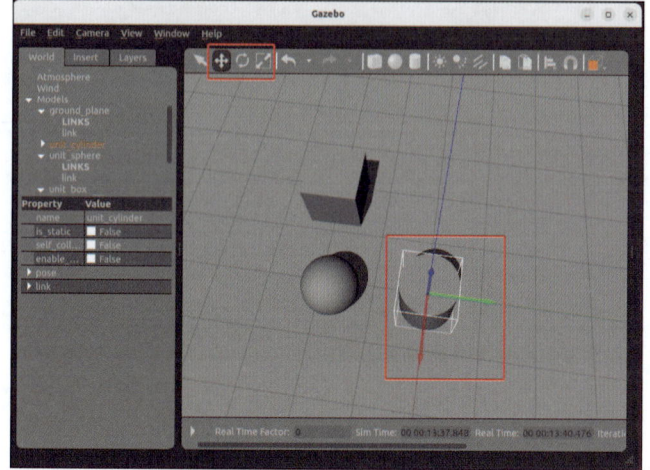

Figure 1-4-5   Transformation model

(2) Edit model

Return to the selection mode, right-click on the model and select Edit model command, as shown in Figure 1-4-6, to enter model editing.

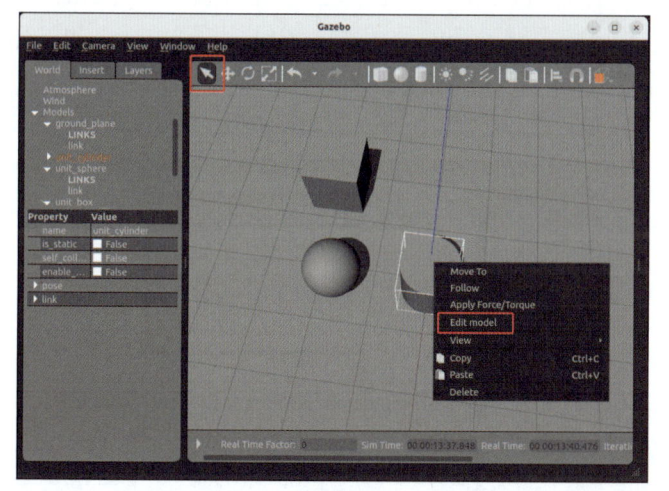

Figure 1-4-6   Select Edit model

Double-click the model in the model editing view, as shown in Figure 1-4-7, to open the model editing dialog box, where you can edit Link, Visual, and Collision in detail.

Click the File menu to save and exit model editing.

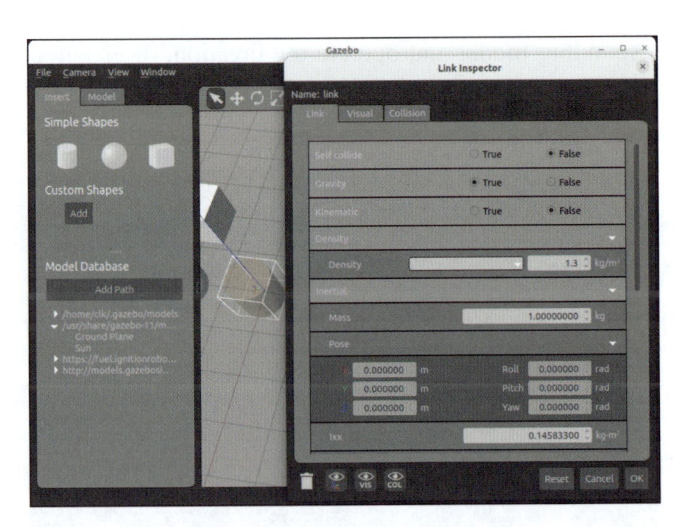

Figure 1-4-7   Model editing

2. Drawing buildings

In the Edit menu, select Building Editor to enter building editing. Gazebo provides Wall, Window, Door and Stairs by default. Let's use Wall to draw a simple wall.

Click the left button of the mouse on the Wall icon, in the white drawing area above the view, click the left button to start drawing, click the left button again to end drawing, repeat the operation, draw multiple walls, click the right button to exit drawing, as shown in Figure 1-4-8.

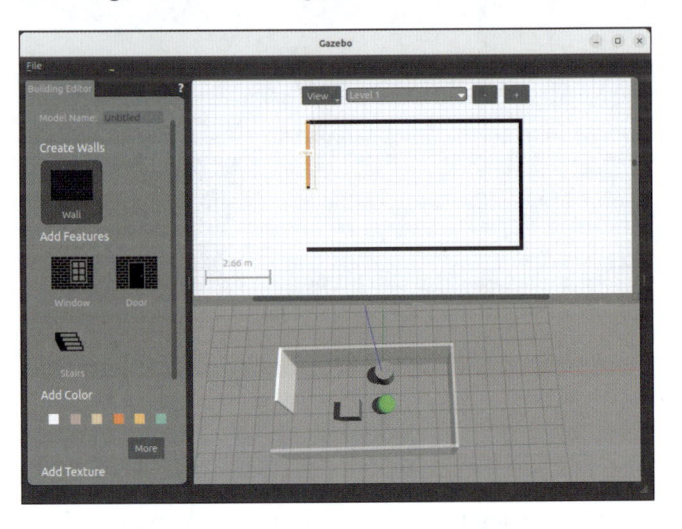

Figure 1-4-8   Drawing a simple wall

Click the File menu to save and exit building editing.

Chapter1   Maze Robot Simulation Foundation   095

Now click the File menu again, select Save World As command, and save the building and model together into Gazebo's world file, for example, named demo.world. Start Gazebo again and open demo. world directly, as shown in Figure 1-4-9.

```
$ gazebo demo.world
```

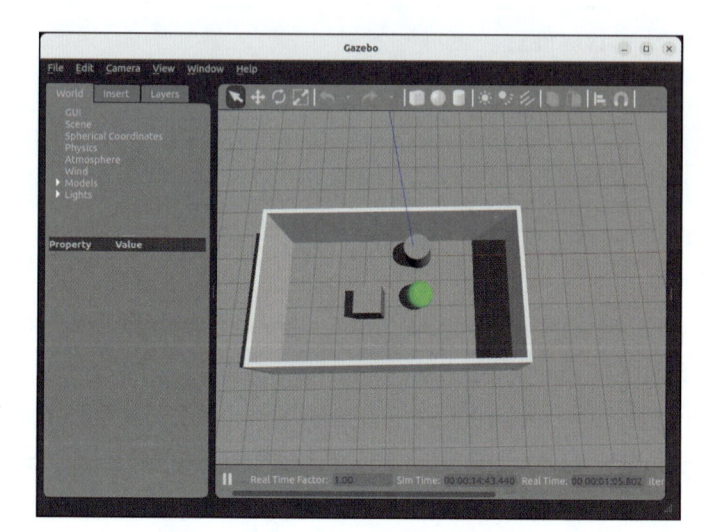

Figure 1-4-9   Open the specified world

Task 2   Learn to Build URDF Model

The purpose of this task is to have a certain understanding of URDF files and learn to write a simple robot URDF file.

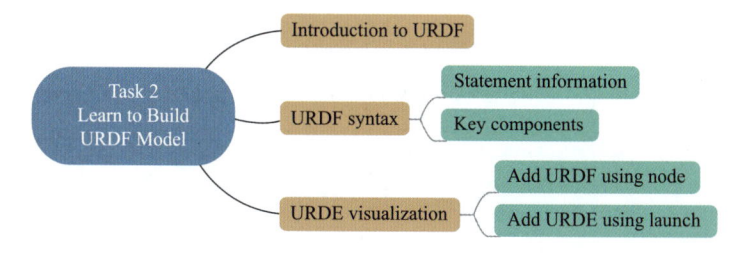

Ⅰ. Introduction to URDF

URDF (unified robot description format) unified robot description format. Similar to the .txt text format and .jpg image format in computer files, URDF is a format based on XML specifications and used to describe the robot structure. It

uses urdf as the file suffix. The purpose of designing this format is to provide a specification for describing robots as general as possible.

ROS official website address, please search it in the search engine yourself.

A simple example, demo. urdf, looks like this:

```xml
<?xml version="1.0"?>
<robot name="mybot">
    <link name="base_link">
        <visual>
            <geometry>
                <cylinder length="0.6" radius="0.2"/>
            </geometry>
        </visual>
    </link>
</robot>
```

From a mechanical perspective, robots are usually modeled as a structure composed of links and joints. A connecting rod is a rigid body with mass attributes, and a joint is a structure that connects and limits the relative motion of two rigid bodies. Connecting the connecting rods in sequence through joints forms a kinematic chain (that is, the robot model defined here). A URDF document describes the relative relationships, inertial properties, geometric characteristics and collision models of a series of joints and connecting rods. Specifically, these include:

① Kinematics and dynamics description of robot model.

② Geometric representation of the robot.

③ Collision model of robot.

II. URDF syntax

A complete URDF consists of declaration information and key components:

```xml
<?xml version="1.0"?>
<robot name="myrobot">
    <link>…</link>
    <link>…</link>
    …
    <joint>…</joint>
    <joint>…</joint>
</robot>
```

For detailed usage of tags, please check the official website.

1. Statement information

There are two types of declaration information, namely XML declaration and robot declaration.

(1) XML declaration

URDF is a description language based on XML specifications, so it needs to be declared as an xml=file, indicating its version number, which is usually located in the first line of the URDF file.

```
<?xml version="1.0"?>
```

(2) Robot statement

Declare everything about the robot.

The robot tag declares a robot model, and the tag pair contains all component information.

```
<robot name="myrobot">
    ...
</robot>
```

2. Key components

There are two main components: link components and joint components.

(1) link component

That is, connecting rods. Each part of the robot is a link. For example, a simple two-wheeled car. The torso, left wheel, right wheel, and support wheel are all a link, as shown in Figure 1-4-10.

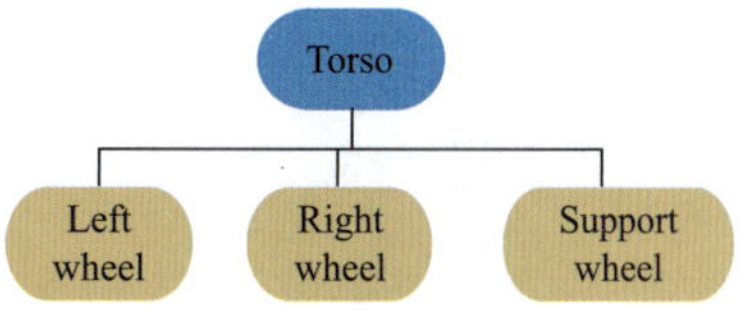

Figure 1-4-10　Schematic diagram of robot links

A link element must contain three parts: visual, collision and inertial. A simple example code is as follows:

```
<!-- base link -->
<link name="base_link">
    <visual>
        <origin xyz="0 0 0.0" rpy="0 0 0" />
        <geometry>
            <cylinder length="0.12" radius="0.10" />
        </geometry>
        <material name="blue">
            <color rgba="0.1 0.1 1.0 0.5" />
        </material>
```

```
        </visual>
        <collision>
            <origin xyz="0 0 0.0" rpy="0 0 0" />
            <geometry>
                <cylinder length="0.12" radius="0.10" />
            </geometry>
            <material name="blue">
                <color rgba="0.1 0.1 1.0 0.5" />
            </material>
        </collision>
        <inertial>
            <mass value="0.2" />
            <inertia ixx="0.0122666" ixy="0" ixz="0" iyy=
"0.0122666" iyz="0" izz="0.02" />
        </inertial>
    </link>
```

(2) joint component

There may be a connection relationship between the links of the robot. This connection is called a joint. For the two-wheeled robot mentioned above, the left wheel, right wheel and support wheel are all connected to the torso through joints, as shown in Figure 1-4-11.

Figure 1-4-11   joint diagram

In the joint, you need to specify the parent link and child link, and determine how they are connected. A simple example code is as follows:

```
<joint name="left_wheel_joint" type="continuous">
    <parent link="base_link" />
      <child link="left_wheel_link" />
    <origin xyz="-0.02 0.10 -0.06" />
    <axis xyz="0 1 0" />
</joint>
```

Ⅲ. URDF visualization

When using Gazebo simulation, use world files to load static resources, such

as houses, roads, mazes, etc., and use URDF files to load dynamic models, such as robots.

Gazebo can directly specify the loaded world file when starting, but the URDF file cannot be loaded directly and needs to be added to Gazebo through the built-in function package of ROS2.

For the full version URDF file, demo. urdf, please scan the QR code to obtain it.

1. Add URDF using node

Start Gazebo and ROS2 plugins.

```
$ gazebo --verbose -s libgazebo_ros_init. so -s libgazebo_ros_
factory. so demo. world
```

If no world file is specified, gazebo will automatically use empty.world in the installation directory (an empty world file).

Open a new terminal and start the function package for adding URDF in ROS2, as shown in Figure 1-4-12.

```
$ ros2 run gazebo_ros spawn_entity. py -entity TQD_Maze robot
-file demo. urdf
```

Video

URDF
visualization

File

Full version
URDF
download

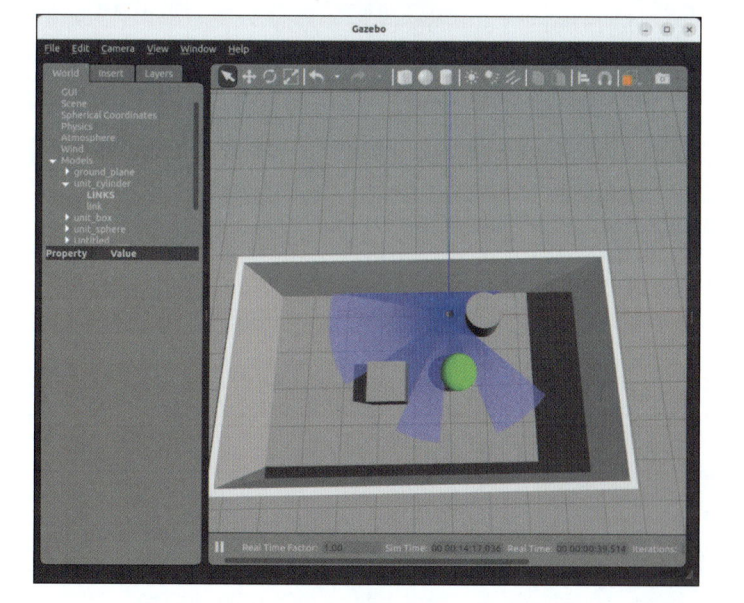

Figure 1-4-12　Add URDF using nodes

gazebo_ros: Function package for adding URDF.

spawn_entity. py: Node file.

-entity TQD_Maze robot : Specify the name of the robot.

-file demo. urdf: Specify the urdf file address.

You can now see the robot in Gazebo.

2. Add URDF using launch

As already introduced in topic communication, launch can start multiple nodes at one time.

Starting Gazebo to load the world and adding URDF files can be treated as starting a node, so it can be completed directly through launch.

Complete the following operations in the pythondemo function package:

Create a new world directory to store world files, and copy demo. world to the world folder.

Create a new urdf directory to store urdf files, and copy demo. urdf to the urdf folder.

(1) Modify launch file

Added to start gazebo and load demo.world directly;

Add the spawn_entity. py node and add demo. urdf.

```
import os
from launch import LaunchDescription
from launch.substitutions import LaunchConfiguration
from launch_ros.actions import Node
from launch.actions import ExecuteProcess
from launch_ros.substitutions import FindPackageShare

def generate_launch_description():
    """
    1. preserved name, ROS2 will recognize these function names
    2. launch content description function, called by ROS2 launch
through scanning
    """
    robot_name_in_model = 'TQD_Robot'
    package_name = 'pythondemo'
    world_name = 'demo.world'
    urdf_name = "demo.urdf"
     pkg_share = FindPackageShare(package=package_name). find
(package_name)
     world_model_path = os.path.join(pkg_share, f'world/{world_
name}')
     urdf_model_path = os.path.join(pkg_share, f'urdf/{urdf_name}')

    # start Gazebo and load world
```

```
        start_gazebo_cmd = ExecuteProcess(
            cmd=['gazebo', '--verbose', '-s', 'libgazebo_ros_init.
so', '-s', 'libgazebo_ros_factory. so', world_model_path],
            output='screen')


        # Add URDF
        spawn_entity_cmd = Node(
            package='gazebo_ros',
            executable='spawn_entity. py',
                arguments=['-entity', robot_name_in_model, '-file',
urdf_model_path], output='screen')
        ld = LaunchDescription()
        ld.add_action(start_gazebo_cmd)
        ld.add_action(spawn_entity_cmd)
        return ld
```

(2) Modify the setup file

Add installation of world files.

Add installation of urdf files.

```
from setuptools import setup
from glob import glob
import os
package_name = 'pythondemo'

setup(
    name=package_name,
    version='0.0.0',
    packages=[package_name],
    data_files=[
        ('share/ament_index/resource_index/packages',
            ['resource/' + package_name]),
        ('share/' + package_name, ['package.xml']),
        (os.path.join('share', package_name, 'launch') , glob
('launch/**')),
    # Install launch file
        (os.path.join('share', package_name, 'world') , glob
('world/**')),
    # Install world file
        (os.path.join('share', package_name, 'urdf') , glob
('urdf/**'))
    # Install urdf file
        ],
    install_requires=['setuptools'],
    zip_safe=True,
    maintainer='clk',
```

```
        maintainer_email='clk@todo.todo',
        description='TODO: Package description',
        license='TODO: License declaration',
        tests_require=['pytest'],
        entry_points={
            'console_scripts': [
        'publisher = pythondemo.publisher:main',
        'subscriber = pythondemo.subscriber:main',
        'demo = pythondemo.demo:main'
            ],
        },
    )
```

(3) Build start

```
$ cd ~/ros2_ws
$ colcon build
$ source install/setup. bash
$ ros2 launch pythondemo pythondemo.launch.py
```

This enables direct loading of world and urdf using launch, as shown in Figure 1-4-13.

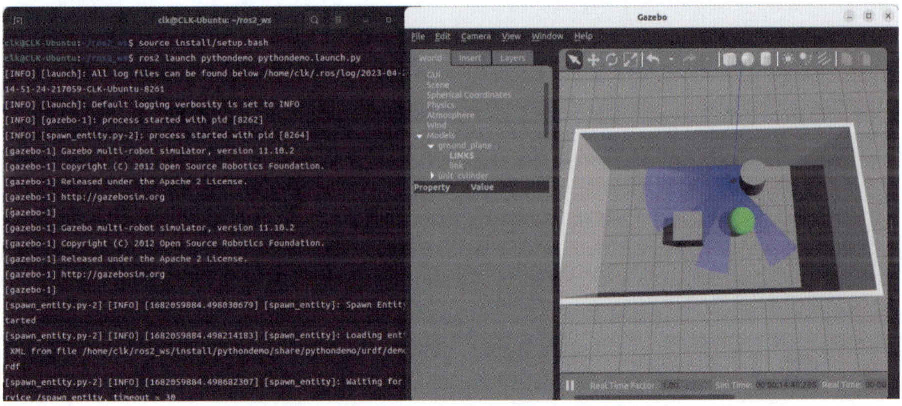

Figure 1-4-13   Add URDF using launch

Questions and Summary

1. How to control the 3D model in Gazebo programmatically?

2. Gazebo can easily load specified maps through command line parameters.

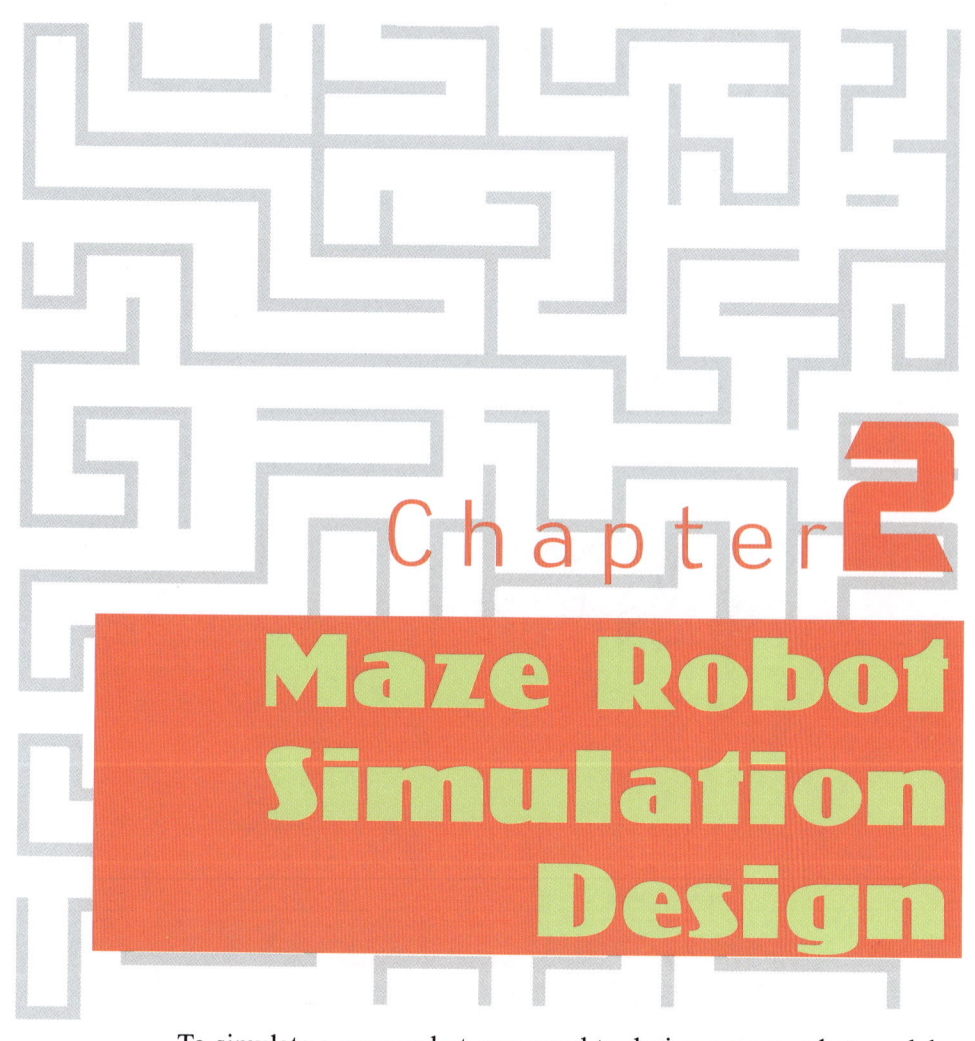

Chapter 2

Maze Robot Simulation Design

To simulate a maze robot, you need to design a maze robot model, then use software for 3D rendering, and finally use a program to control it. In this chapter, we mainly explain how to make a maze robot model, render it in Gazebo, and control the model to achieve various actions through ROS communication.

Chapter2　Maze Robot Simulation Design　105

Project 1

Learn How to Make a Maze Robot

Learning objectives

1. Master the method of making a virtual simulation maze robot according to international standards.

2. Master the method of creating a virtual simulation maze according to international standards.

Task 1　Make a Virtual Simulation Maze Robot

The purpose of this task is to learn to make your own virtual simulation maze robot.

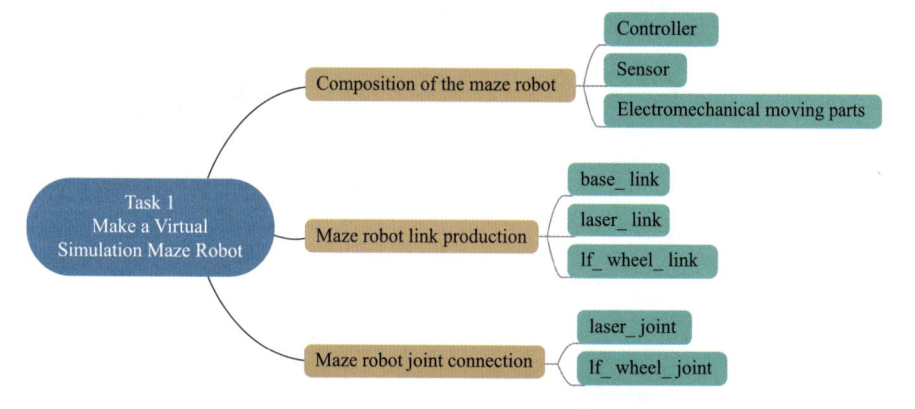

Ⅰ. Composition of the maze robot

The real maze robot consists of three parts, the controller, the sensor and the electromechanical moving parts, as shown in Figure 2-1-1. The basic function is to run from the starting point to the end point. The sensor detects the surrounding wall information during operation, and the controller processes the data and combines it with its own algorithm to control the operation of the

electromechanical moving parts.

The virtual simulation maze robot is a virtual maze robot simulated by ROS2 and Gazebo based on the URDF model file, and also includes these three parts.

Figure 2-1-1   Real maze robot

1. Controller

ROS2 topics and messages, ROS2 reserved topic communication API, it can achieve the control of the virtual simulation maze robot.

2. Sensor

Gazebo plug-in, lidar plug-in, can realize the function of infrared sensor and detect the distance of surrounding obstacles.

3. Electromechanical moving parts

Gazebo plug-in, a differential drive plug-in, can realize the function of the motor and drive the model to go straight and turn.

Therefore, the virtual simulation maze robot requires the cooperation of ROS2 and Gazebo to operate normally.

In the following content, unless otherwise specified, the maze robot refers to the virtual simulation maze robot.

A simple example of a four-wheel maze robot is shown in Figure 2-1-2.

The maze robot contains a total of six links, as shown in Figure 2-1-3.

Figure 2-1-2   Example of four-wheel maze robot

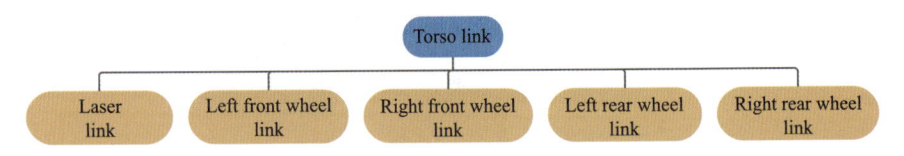

Figure 2-1-3   Six links

The torso is the main link of the maze robot and is connected to five sub-links. The laser link is used to implement the lidar plug-in in Gazebo. The four wheels are the motion structure of the maze robot. The sub-link and the main link are connected through joints to form a complete maze robot.

Ⅱ. Maze robot link production

Referring to international competition rules, the maze robot has size restrictions when running in the maze. If it is too large or too small, it will have an inconvenient impact on the operation.

The maze cell size is 18 cm × 18 cm, and the baffle size is 18 cm × 1.2 cm × 5 cm. Therefore, the recommended maze robot size is 9 cm long, 8 cm wide, and 3 cm high.

Maze robot link production reference program:

1. base_link

```xml
<link name="base_link">
    <visual>   <!-- visual elements, describing the shape, size
and color of the link, etc. -->
        <origin rpy="0 0 0" xyz="0 0 0" />
        <geometry>
            <box size="0.09 0.06 0.01" />
        </geometry>
        <material name="blue">
            <color rgba="0.0 0.0 0.8 1.0" />
        </material>
    </visual>
    <collision> <!-- collision elements, references that describe
collisions with other models -->
        <origin rpy="0 0 0" xyz="0 0 0" />
        <geometry>
            <box size="0.09 0.06 0.01" />
        </geometry>
        <material name="green">
            <color rgba="0.0 0.8 0.0 1.0" />
        </material>
    </collision>
    <inertial> <!-- inertial element, describing the link's
mass and rotational inertia matrix, etc. -->
        <origin rpy="1.5707963267948966 0 1.5707963267948966"
xyz="0 0 0" />
        <mass value="0.4" />
        <inertia ixx="0.0001233333333333333" ixy="0.0" ixz=
"0.0" iyy="0.00038999999999999994"
            iyz="0.0" izz="0.0002733333333333333" />
    </inertial>
</link>
```

2. laser_link

```xml
<link name="laser_link">
    <visual>   <!-- visual elements, describing the shape, size
and color of the link, etc. -->
        <origin rpy="0 0 0" xyz="0 0 0" />
        <geometry>
```

```xml
                <cylinder length="0.01" radius="0.005" />
            </geometry>
            <material name="blue">
                <color rgba="0.0 0.0 0.8 1.0" />
            </material>
        </visual>
        <collision>  <!-- collision elements, references that describe
collisions with other models -->
            <origin rpy="0 0 0" xyz="0 0 0" />
            <geometry>
                <cylinder length="0.01" radius="0.005" />
            </geometry>
            <material name="blue">
                <color rgba="0.0 0.0 0.8 1.0" />
            </material>
        </collision>
        <inertial> <!-- inertial element, describing the link's mass
and rotational inertia matrix, etc. -->
            <origin rpy="1.5707963267948966 0 0" xyz="0 0 0" />
            <mass value="0.1" />
            <inertia ixx="1.4583333333333333e-06" ixy="0" ixz=
"0" iyy="1.4583333333333333e-06"
                iyz="0" izz="1.25e-06" />
        </inertial>
    </link>
```

3. lf_wheel_link

The four-wheel links are exactly the same. Please note that they are named with different names, such as rf_wheel_link, lr_wheel_link and rr_wheel_link.

```xml
        <link name="lf_wheel_link">
            <visual>   <!-- visual elements, describing the shape, size
and color of the link, etc. -->
                <origin rpy="1.57079 0 0" xyz="0 0 0" />
                <geometry>
                    <cylinder length="0.01" radius="0.01" />
                </geometry>
                <material name="blue">
                    <color rgba="0.0 0.0 0.8 1.0" />
                </material>
            </visual>
            <collision> <!-- collision elements, references that describe
collisions with other models -->
                <origin rpy="1.57079 0 0" xyz="0 0 0" />
                <geometry>
                    <cylinder length="0.01" radius="0.01" />
                </geometry>
                <material name="blue">
                    <color rgba="0.0 0.0 0.8 1.0" />
                </material>
            </collision>
```

```
        <inertial> <!-- inertial element, describing the link's
mass and rotational inertia matrix, etc. -->
            <origin rpy="1.5707963267948966 0 0" xyz="0 0 0" />
            <mass value="0.1" />
            <inertia ixx="3.333333333333333e-06" ixy="0" ixz="0"
iyy="3. 333333333333333e-06" iyz="0"
                izz="5e-06" />
        </inertial>
    </link>
```

The link describes the appearance, collision and inertial properties of the maze robot. If the parameters do not match or conflict, it will have a huge impact on the movement of the maze robot. Therefore, it is recommended to use the TQD evaluation system to quickly generate the maze robot link with one click.

III. Maze robot joint connection

The connection position and method of the sub-link and the main link determine the type of the joint.

Maze robot joint connection reference program:

1. laser_joint

```
        <joint name="laser_joint" type="fixed"> <!-- laser link
does not require rotation, fixed -->
            <parent link="base_link" />
            <child link="laser_link" />
            <origin xyz="0.025 0 0.01" />
        </joint>
```

2. lf_wheel_joint

The four wheels are basically the same as the joints of the main link. The only thing that needs attention is that the location of the connection is different.

```
        <joint name="lf_wheel_joint" type="continuous"> <!-- the
wheel link can rotate infinitely, continuous -->
            <!-- left front-->
            <parent link="base_link" />
            <child link="lf_wheel_link" />
            <origin xyz="0.005 0.035 0" />
            <axis xyz="0 1 0" />
        </joint>
```

After completing the link and joint, a maze robot is ready. Now it does not have sensors (lidar) and electromechanical moving parts (differential drive). This book will introduce it in Project 2 of this Chapter.

The complete maze robot URDF file can be obtained by scanning the QR code

File

Maze robot URDF file

Task 2　Create a Virtual Simulation Maze

The purpose of this task is to learn to create a virtual simulation maze.

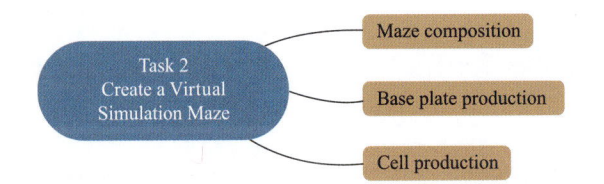

Ⅰ. Maze composition

A real maze that complies with IEEE international standards is shown in Figure 2-1-4.

It contains a total of 16 × 16 cells, each cell is 18 cm × 18 cm in size. Every change of the wall will form a new maze.

The virtual simulation maze is the same as the real maze, including the base plate and the baffle. Since there is no need to manually insert and remove the baffle, the columns are merged into the baffle in the virtual simulation maze, as shown in Figure 2-1-5.

Figure 2-1-4　Real maze　　　　Figure 2-1-5　Example of virtual maze

The virtual simulation maze floor is a whole and can be treated directly as a rectangular parallelepiped. The cell walls are independent. It is recommended to initialize all the walls above the first row of the maze and the walls to the right of the last column. For each cell, only the walls to the left and below are considered.

Ⅱ. Base plate production

The base plate is a link, which needs to contain three parts: visual, collision and inertial.

```xml
<visual name="visual">
    <pose frame="">0 0 0.0025 0 -0 0</pose>
    <geometry>
        <box>
            <size>2.892 2.892 0.005</size>
        </box>
    </geometry>
    <material>
        <script>
            <uri>file://media/materials/scripts/gazebo. material</uri>
            <name>Gazebo/White</name>
        </script>
    </material>
</visual>
<collision name="collision">
    <pose frame="">0 0 0.0025 0 -0 0</pose>
    <geometry>
        <box>
            <size>2.892 2.892 0.005</size>
        </box>
    </geometry>
    <surface>
        <friction>
            <ode>
                <mu>1000</mu>
                <mu2>1000</mu2>
            </ode>
            <torsional>
                <ode />
            </torsional>
        </friction>
        <contact>
            <ode />
        </contact>
        <bounce />
    </surface>
    <max_contacts>10</max_contacts>
</collision>
<inertial>
    <mass>1</mass>
    <inertia>
        <ixx>1.39394</ixx>
        <ixy>0</ixy>
        <ixz>0</ixz>
        <iyy>0.696972</iyy>
        <iyz>0</iyz>
        <izz>0.696973</izz>
    </inertia>
    <pose frame="">0 0 0.05 0 -0 0</pose>
</inertial>
```

III. Cell production

Each cell contains four baffles. After the walls above the first row of the maze and the walls on the right side of the last column are all initialized and drawn, only the walls to the left and below are considered for each cell.

Draw the north wall:

```
<visual name="paint_v_0_0_N">
    <pose frame="">-1.35 1.44 0.05 0 -0 0</pose>
    <geometry>
        <box>
            <size>0.192 0.012 0.01</size>
        </box>
    </geometry>
    <material>
        <script>
            <name>Gazebo/Red</name>
            <uri>file://media/materials/scripts/gazebo. material</uri>
        </script>
    </material>
</visual>
<visual name="v_0_0_N">
    <pose frame="">-1.35 1.44 0.025 0 -0 0</pose>
    <geometry>
        <box>
            <size>0.192 0.012 0.04</size>
        </box>
    </geometry>
</visual>
<collision name="p_0_0_N">
    <pose frame="">-1.35 1.44 0.03 0 -0 0</pose>
    <geometry>
        <box>
            <size>0.192 0.012 0.05</size>
        </box>
    </geometry>
    <max_contacts>10</max_contacts>
    <surface>
        <contact>
            <ode />
        </contact>
        <bounce />
        <friction>
            <torsional>
                <ode />
            </torsional>
            <ode />
        </friction>
    </surface>
</collision>
```

Chapter2 Maze Robot Simulation Design | 113

Draw the east wall:

```
<visual name="paint_v_0_15_E">
    <pose frame="">1.44 1.35 0.05 0 -0 1.5708</pose>
    <geometry>
        <box>
            <size>0.192 0.012 0.01</size>
        </box>
    </geometry>
    <material>
        <script>
            <name>Gazebo/Red</name>
            <uri>file://media/materials/scripts/gazebo.material</uri>
        </script>
    </material>
</visual>
<visual name="v_0_15_E">
    <pose frame="">1.44 1.35 0.025 0 -0 1.5708</pose>
    <geometry>
        <box>
            <size>0.192 0.012 0.04</size>
        </box>
    </geometry>
</visual>
<collision name="p_0_15_E">
    <pose frame="">1.44 1.35 0.03 0 -0 1.5708</pose>
    <geometry>
        <box>
            <size>0.192 0.012 0.05</size>
        </box>
    </geometry>
    <max_contacts>10</max_contacts>
    <surface>
        <contact>
            <ode />
        </contact>
        <bounce />
        <friction>
            <torsional>
                <ode />
            </torsional>
            <ode />
        </friction>
    </surface>
</collision>
```

Draw the west wall:

```
<visual name="paint_v_0_0_W">
    <pose frame="">-1.44 1.35 0.05 0 -0 1.5708</pose>
    <geometry>
```

```
            <box>
                <size>0.192 0.012 0.01</size>
            </box>
        </geometry>
        <material>
            <script>
                <name>Gazebo/Red</name>
                <uri>file://media/materials/scripts/gazebo.material</uri>
            </script>
        </material>
    </visual>
    <visual name="v_0_0_W">
        <pose frame="">-1.44 1.35 0.025 0 -0 1.5708</pose>
        <geometry>
            <box>
                <size>0.192 0.012 0.04</size>
            </box>
        </geometry>
    </visual>
    <collision name="p_0_0_W">
        <pose frame="">-1.44 1.35 0.03 0 -0 1.5708</pose>
        <geometry>
            <box>
                <size>0.192 0.012 0.05</size>
            </box>
        </geometry>
        <max_contacts>10</max_contacts>
        <surface>
            <contact>
                <ode></ode>
            </contact>
            <bounce></bounce>
            <friction>
                <torsional>
                    <ode></ode>
                </torsional>
                <ode></ode>
            </friction>
        </surface>
    </collision>
```

Draw the southern wall:

```
Draw the southern wall. <visual name="paint_v_0_1_S">
    <pose frame="">-1.17 1.26 0.05 0 -0 0</pose>
    <geometry>
        <box>
            <size>0.192 0.012 0.01</size>
        </box>
    </geometry>
    <material>
        <script>
```

Chapter2　Maze Robot Simulation Design | 115

```xml
            <name>Gazebo/Red</name>
            <uri>file://media/materials/scripts/gazebo.material</uri>
        </script>
    </material>
</visual>
<visual name="v_0_1_S">
    <pose frame="">-1.17 1.26 0.025 0 -0 0</pose>
    <geometry>
        <box>
            <size>0.192 0.012 0.04</size>
        </box>
    </geometry>
</visual>
<collision name="p_0_1_S">
    <pose frame="">-1.17 1.26 0.03 0 -0 0</pose>
    <geometry>
        <box>
            <size>0.192 0.012 0.05</size>
        </box>
    </geometry>
    <max_contacts>10</max_contacts>
    <surface>
        <contact>
            <ode></ode>
        </contact>
        <bounce></bounce>
        <friction>
            <torsional>
                <ode></ode>
            </torsional>
            <ode></ode>
        </friction>
    </surface>
</collision>
```

Since the maze has a total of 256 cells, it is difficult to write the code manually. Therefore, it is recommended to use the TQD evaluation system to quickly generate the maze with one click based on the TQD model file. Once you've finished creating your maze, you can display it in Gazebo.

The complete maze world file can be obtained by scanning the QR code

Task 3　Learning Virtual Simulation System Startup

The purpose of this task is to use the ROS2 function package to start the world file and urdf file produced by yourself, and lay the foundation for the future development of the maze robot.

File

Maze
world file

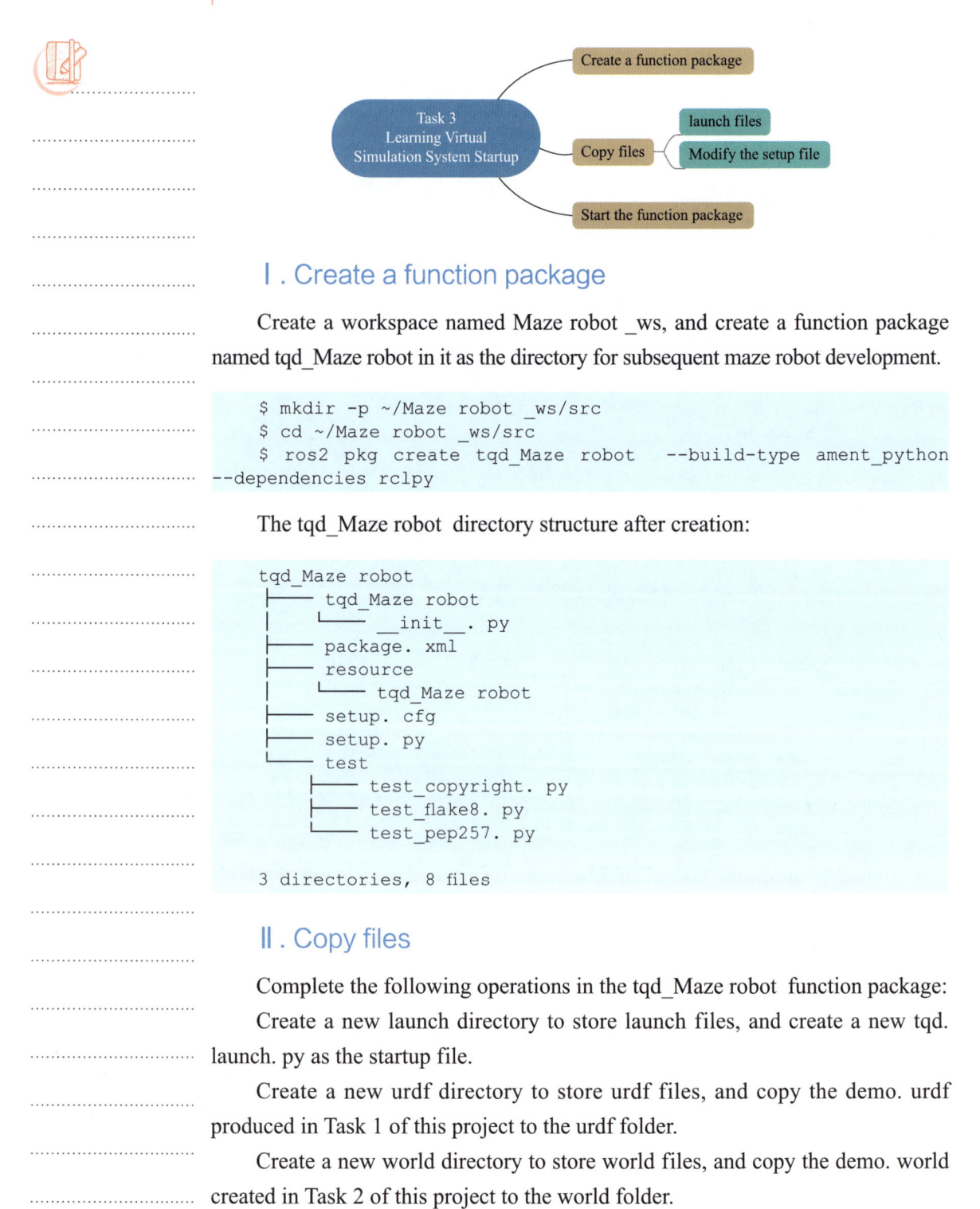

I. Create a function package

Create a workspace named Maze robot _ws, and create a function package named tqd_Maze robot in it as the directory for subsequent maze robot development.

```
$ mkdir -p ~/Maze robot _ws/src
$ cd ~/Maze robot _ws/src
$ ros2 pkg create tqd_Maze robot  --build-type ament_python
--dependencies rclpy
```

The tqd_Maze robot directory structure after creation:

```
tqd_Maze robot
├──── tqd_Maze robot
│     └──── __init__. py
├──── package. xml
├──── resource
│     └──── tqd_Maze robot
├──── setup. cfg
├──── setup. py
└──── test
      ├──── test_copyright. py
      ├──── test_flake8. py
      └──── test_pep257. py

3 directories, 8 files
```

II. Copy files

Complete the following operations in the tqd_Maze robot function package:

Create a new launch directory to store launch files, and create a new tqd. launch. py as the startup file.

Create a new urdf directory to store urdf files, and copy the demo. urdf produced in Task 1 of this project to the urdf folder.

Create a new world directory to store world files, and copy the demo. world created in Task 2 of this project to the world folder.

Chapter2　Maze Robot Simulation Design 117

1. launch file

```
from launch import LaunchDescription
from launch_ros.actions import Node
from launch.actions import ExecuteProcess
from launch_ros.substitutions import FindPackageShare
import os

def generate_launch_description():
    robot_name_in_model = 'TQD_Robot'
    package_name = 'tqd_Maze robot'
    urdf_name = 'demo.urdf'
    world_name = 'demo.world'

    ld = LaunchDescription()
     pkg_share = FindPackageShare(package=package_name).find
(package_name)
      urdf_model_path = os.path.join(pkg_share,f'urdf/{urdf_
name}')
      world_model_path = os.path.join(pkg_share,f'world/{world_
name}')

    # Start Gazebo server
    start_gazebo_cmd = ExecuteProcess(
        cmd=['gazebo', '--verbose', '-s', 'libgazebo_ros_factory.
so', world_model_path],  # , world_model_path
        output='screen'
    )

    # Launch the robot
    spawn_entity_cmd = Node(
        package='gazebo_ros',
        executable='spawn_entity.py',
        arguments=['-entity', robot_name_in_model,  '-file', urdf_
model_path, '-x', '-1.35', '-y', '-1.35', '-z', '0.10', '-Y', '1.575'],
        output='screen'
    )

    ld = LaunchDescription()
    ld.add_action(start_gazebo_cmd)
    ld.add_action(spawn_entity_cmd)
    return ld
```

2. Modify the setup file

```
from setuptools import setup
from glob import glob
import os

package_name = 'tqd_Maze robot'

setup(
```

```
        name=package_name,
        version='0.0.0',
        packages=[package_name],
        data_files=[
            ('share/ament_index/resource_index/packages',
                ['resource/' + package_name]),
            ('share/' + package_name, ['package.xml']),
            (os.path.join('share', package_name, 'launch'),
glob('launch/*.launch.py')),
            (os.path.join('share', package_name, 'urdf'),
glob('urdf/**')),
            (os.path.join('share', package_name, 'world'),
glob('world/**')),
        ],
        install_requires=['setuptools'],
        zip_safe=True,
        maintainer='clk',
        maintainer_email='clk@todo.todo',
        description='TODO: Package description',
        license='TODO: License declaration',
        tests_require=['pytest'],
        entry_points={
            'console_scripts': [
            ],
        },
    )
```

3. Start the function package

```
$ cd ~/Maze robot _ws
$ colcon build
$ source install/setup.bash
$ ros2 launch tqd_Maze robot  tqd.launch.py
```

Start world and urdf as shown in Figure 2-1-6.

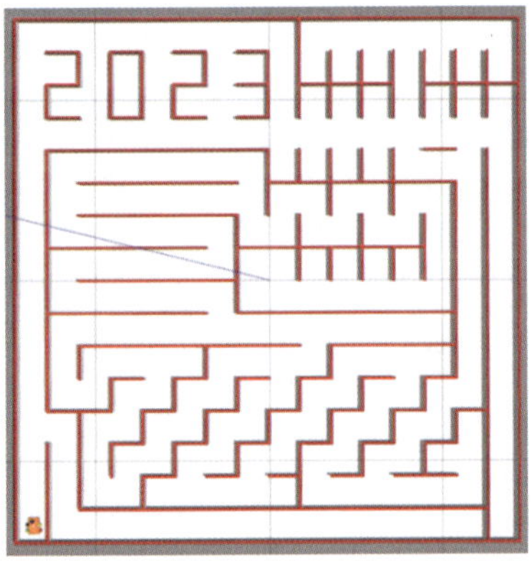

Questions and Summary

1. When making a maze robot, how should the four wheel be placed?

2. How to accurately position each cell when drawing the maze?

3. The maze and urdf in the world are two different models. You should pay attention to the relative positions of the two when displaying them.

Figure 2-1-6   Start world and urdf at the same time

Chapter2 Maze Robot Simulation Design | 119

Project 2

Understand the Basic Functions of a Maze Robot

Learning objectives

1. Master the principles and usage of two-wheel differential plug-in.
2. Master the principles and usage of lidar plug-in.
3. Master multi-topic collaborative work program control.

Task 1 Understand Two-wheel Differential Drive

The purpose of this task is to learn to use the two-wheel differential plug-in to drive the maze robot.

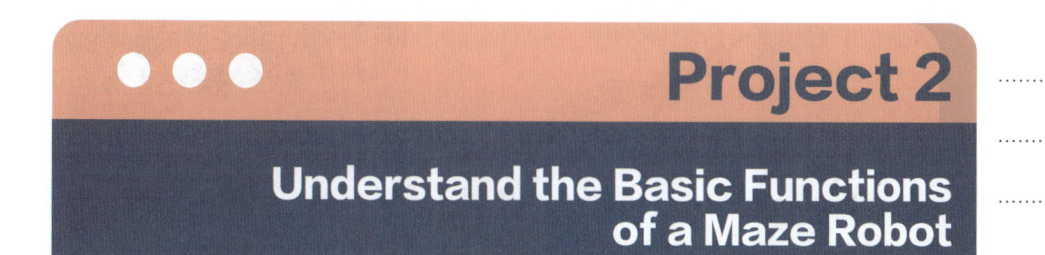

Ⅰ. Introduction to the two-wheel differential plug-in

The driving methods of the robot mainly include motor driving, hydraulic driving, and pneumatic driving, as shown in Figure 2-2-1.

Motor driver is to use the force or torque generated by various motors to directly drive the joints of the robot, or to drive the joints of the robot through a mechanism such as deceleration, to obtain the required position, speed, acceleration and other indicators. It has the advantages of environmental protection, cleanliness, convenient control, high movement precision, low maintenance cost and high driving efficiency.

(a) Motor drive mode　　　(b) Pneumatic drive mode　　　(c) Hydraulic drive mode

Figure 2-2-1　Three driving modes of the robot

Hydraulic actuators use liquid as a medium to transmit force, and use a hydraulic pump to make the pressure generated by the hydraulic system drive the actuator movement. The hydraulic drive mode is a mature drive mode.

Pneumatic drives use air as the working medium and use an air source generator to convert the pressure energy of compressed air into mechanical energy to drive the actuator to complete the predetermined motion law. Pneumatic drive has the advantages of simple energy saving, short time, fast action, softness, light weight, high output/quality ratio, convenient installation and maintenance, safety, low cost, and no pollution to the environment.

The maze robot adopts a wheeled structure and uses a two-wheel differential plug-in to simulate motor drive.

The two-wheel differential plug-in is a plug-in in Gazebo used to control the movement of the model. The dynamic link library gazebo_ros_diff_drive.so can receive commands to set the speed of the model, and can also feedback the position and speed of the model, as shown in Figure 2-2-2.

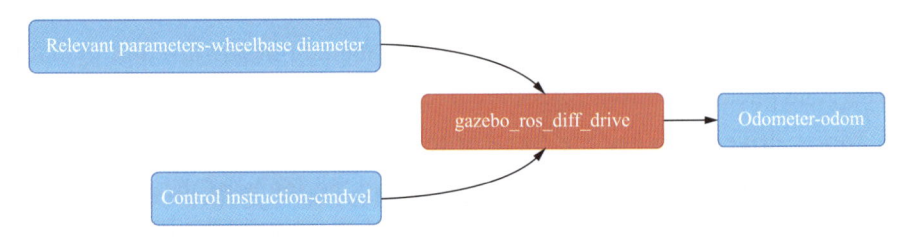

Figure 2-2-2　Two-wheel differential plug-in

The prerequisite for the two-wheel differential plug-in to take effect is to correctly import the plug-in in urdf and configure the relevant parameters.

Chapter2   Maze Robot Simulation Design | 121

By default, the two-wheel differential plug-in obtains the target linear speed and target angular speed by subscribing to the topic /cmd_vel. The message type of this topic is geometry_msgs/msg/Twist. You can publish data to this topic to control the target linear speed and target angular speed.

By default, the two-wheel differential plug-in outputs the current position, orientation (yaw angle) and speed of the target by publishing the topic /odom. The message type of this topic is nav_msgs/msg/Odometry. You can subscribe to this topic to obtain the current position, heading (yaw angle) and speed of the target.

II . Adding two-wheel differential plug-in

Add the two-wheel differential code in urdf and configure related parameters.

```
    <gazebo>
        <plugin name='diff_drive' filename='libgazebo_ros_diff_
drive.so'>
            <ros>
                <namespace>/</namespace>    <!-- namespace-->
                    <remapping>cmd_vel:=cmd_vel</remapping> <!--
remap topic names-->
                    <remapping>odom:=odom</remapping> <!-- remap
topic names-->
            </ros>
            <update_rate>60</update_rate>    <!-- update frequency-->
            <!-- wheels -->
            <left_joint>left_wheel_joint</left_joint> <!-- left
joint-->
             <right_joint>right_wheel_joint</right_joint> <!--
right joint-->
            <!-- kinematics -->
            <wheel_separation>0.07</wheel_separation> <!-- center
distance between two wheels, unit m-->
            <wheel_diameter>0.02</wheel_diameter> <!-- diameter
of wheel, unit m-->
            <!-- limits -->
            <max_wheel_torque>20</max_wheel_torque> <!-- torque
of wheel, unit Nm-->
             <max_wheel_acceleration>1. 0</max_wheel_acceleration>
<!-- acceleration of wheel-->
            <!-- output -->
            <publish_odom>true</publish_odom>    <!-- whether
to publish odom topics-->
            <publish_odom_tf>true</publish_odom_tf>
            <publish_wheel_tf>false</publish_wheel_tf>
```

```
            <odometry_frame>odom</odometry_frame>
            <robot_base_frame>base_footprint</robot_base_frame>
        </plugin>
    </gazebo>
```

Ⅲ. Topic of two-wheel differential speed

After adding the two-wheel differential code, rebuild and launch tqd. launch. py

```
$ cd ~/Maze robot _ws
$ colcon build
$ source install/setup. bash
$ ros2 launch tqd_Maze robot  tqd. launch. py
```

Print the topic list.

```
$ ros2 topic list
```

The following information is output in the terminal:

```
/camera_image
/camera_info
/clock
/cmd_vel          # set target linear speed and target angular speed
/joint_states
/odom             # output the current position, heading (yaw angle)
                  # and speed of the target
/parameter_events
/rosout
/tf
```

1. /cmd_vel topic

(1) Print topic info

```
$ ros2 topic info /cmd_vel
```

The following information is output in the terminal:

```
Type: geometry_msgs/msg/Twist
Publisher count: 0
Subscription count: 1
```

There is no publisher yet, so the Publisher count is 0.

(2) Print topic echo

Since the plug-in subscribes to the /cmd_vel topic, there are no echo messages for this topic without a publisher.

Chapter2   Maze Robot Simulation Design   123

(3) View message type

```
$ ros2 interface show geometry_msgs/msg/Twist
```

The following information is output in the terminal:

```
# This expresses velocity in free space broken into its linear and
angular parts.

Vector3  linear
    float64 x
    float64 y
    float64 z
Vector3  angular
    float64 x
    float64 y
    float64 z
```

2. /odom topic

(1) Print topic info

```
$ ros2 topic info /odom
```

The following information is output in the terminal:

```
Type: sensor_msgs/msg/Odometry
Publisher count: 1
Subscription count: 0
```

There are no subscribers yet, so the Subscription count is 0.

(2) Print topic echo

```
$ ros2 topic echo /odom
```

The following information is output in the terminal:

```
header:
  stamp:                          # published date
    sec: 1459
    nanosec: 189000000
  frame_id: odom
child_frame_id: base_footprint
pose:                             # pose information
  pose:
    position:                     # position information
      x: 0.0012131783619373808
      y: -2.584250942327375e-05
      z: 0.015999687192963755
```

```
      orientation:                       # orientation information
        x: 3.452774919449117e-07
        y: 7.119953618926764e-06
        z: -0.0005609542394105853
        w: 0.999999842639752
    covariance:                           # matrix
    - 1.0e-05
    - 0.0
    - 0.0
    - 0.0
    - 0.0
    - 0.0
    - 0.0
    - 1.0e-05
    - 0.0
    - 0.0
    - 0.0
    - 0.0
    - 0.0
    - 0.0
    - 1000000000000.0
    - 0.0
    - 0.0
    - 0.0
    - 0.0
    - 0.0
    - 1000000000000.0
    - 0.0
    - 0.0
    - 0.0
    - 0.0
    - 0.0
    - 1000000000000.0
    - 0.0
    - 0.0
    - 0.0
    - 0.0
    - 0.0
    - 0.001
  twist:                                  # current speed
    twist:
      linear:                             # current line speed
        x: -6.892120916898483e-07
        y: -1.2969665645975879e-07
```

```
         z: 0.0
      angular:                              # current angular speed
         x: 0.0
         y: 0.0
         z: -6.453089008299199e-06
   covariance:
   - 1.0e-05
   - 0.0
   - 0.0
   - 0.0
   - 0.0
   - 0.0
   - 1.0e-05
   - 0.0
   - 0.0
   - 0.0
   - 0.0
   - 0.0
   - 0.0
   - 1000000000000.0
   - 0.0
   - 0.0
   - 0.0
   - 0.0
   - 0.0
   - 1000000000000.0
   - 0.0
   - 0.0
   - 0.0
   - 0.0
   - 0.0
   - 1000000000000.0
   - 0.0
   - 0.0
   - 0.0
   - 0.0
   - 0.0
   - 0.001
```

(3) Print message type

```
$ ros2 interface show nav_msgs/msg/Odometry
```

The following information is output in the terminal:

```
# This represents an estimate of a position and velocity in free space.
# The pose in this message should be specified in the coordinate
# frame given by header. frame_id
# The twist in this message should be specified in the coordinate
# frame given by the child_frame_id

# Includes the frame id of the pose parent.
std_msgs/Header header
    builtin_interfaces/Time stamp
            int32 sec
            uint32 nanosec
    string frame_id

# Frame id the pose points to. The twist is in this coordinate frame.
# string child_frame_id

# Estimated pose that is typically relative to a fixed world frame.
# geometry_msgs/PoseWithCovariance pose
    Pose pose
            Point position
                    float64 x
                    float64 y
                    float64 z
            Quaternion orientation
                    float64 x 0
                    float64 y 0
                    float64 z 0
                    float64 w 1
    float64[36] covariance

# Estimated linear and angular velocity relative to child_frame_id.
# geometry_msgs/TwistWithCovariance twist
    Twist twist
            Vector3  linear
                    float64 x
                    float64 y
                    float64 z
            Vector3  angular
                    float64 x
                    float64 y
                    float64 z
    float64[36] covariance
```

IV. Two-wheel differential program control

The following uses a Python program to interact with the differential plug-in. Since the Python program is started manually externally, there is no need to create a node in tqd_maze robot. Create a scripts directory directly in the home directory to place the practice script file.

• Video

Two-wheel
differential
drive

Chapter2   Maze Robot Simulation Design   127

1. /cmd_vel topic

Create diff_drive_cmd_vel. py in the scripts directory, use a Python program to publish data to the /cmd_vel topic, and set the target speed for the maze robot.

```
import rclpy
from rclpy. node import Node
from geometry_msgs. msg import Twist


def set_speed():
    rclpy. init(args=None)            # initialization
node = Node('publisher')             # create node
#create publisher, topic/cmd_vel
    publisher = node. create_publisher(Twist, '/cmd_vel', 10)
    msg = Twist()                    # message type

    def set_speed_callback():
        msg.linear.x = 0.1           # x linear speed assignment
        msg.angular.z = 0.0          # z angular velocity assignment
        publisher. publish(msg)

    node. create_timer(0.5, set_speed_callback)
    rclpy. spin(node)  # execute the callback. After executing this
                       # program, ROS2 will start working
    rclpy. shutdown()  # when the program ends (for example, Ctrl+C),
                       # close the ROS2 node

set_speed()
```

After running the program, the maze robot will walk along a straight line, as shown in Figure 2-2-3. Try to modify msg. linear. x and msg. angular. z and observe the running status and trajectory.

2. /odom topic

Create diff_drive_odom. py in the scripts directory, use a Python program to subscribe to the /odom topic, and obtain the position and yaw angle of the maze robot in Gazebo.

In ROS1, the tf module for coordinate

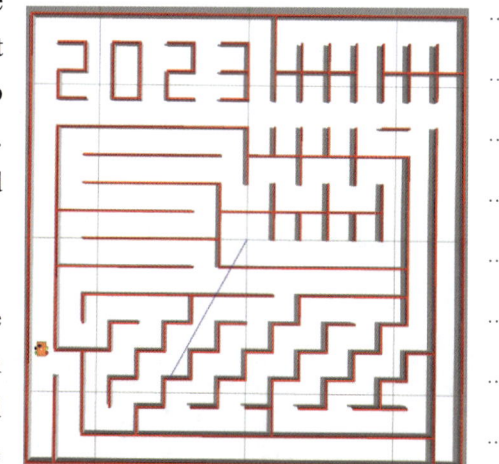

Figure 2-2-3   Driving the movement of the maze robot

conversion is integrated, while in ROS2, you need to install it yourself:

```
$ sudo apt install python3-pip      # install pip to facilitate
                                    # management of python packages
$ pip install transforms3d
$ sudo apt install ros-humble-tf-transformations


import rclpy
from rclpy. node import Node
from nav_msgs.msg import Odometry
import tf_transformations


def get_odom_callback(msg):
    position = msg.pose.pose.position
    quaternion = (
        msg.pose.pose.orientation.x,
        msg.pose.pose.orientation.y,
        msg.pose.pose.orientation.z,
        msg.pose.pose.orientation.w)
    euler = tf_transformations. euler_from_quaternion(quaternion)
                                    #convert quaternions to Euler
                                    #angles
    print(position. x, position. y) #the coordinates of the robot
                                    #in Gazebo
    print(euler[2])                 #yaw angle, -3.14 - 3.14

def get_odom():
    rclpy.init(args=None)
node = Node('subscriber')
#create subscriber, topic/odom, callback get_odom_callback
    node.create_subscription(Odometry, '/odom', get_odom_callback, 10)
    rclpy.spin(node)
    rclpy.shutdown()

get_odom()
```

Run the program, move or rotate the maze robot, and observe the output data, as shown in Figure 2-2-4 and Figure 2-2-5.

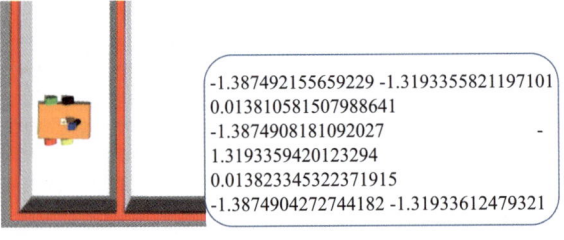

Figure 2-2-4   Position 1

Chapter2　Maze Robot Simulation Design　129

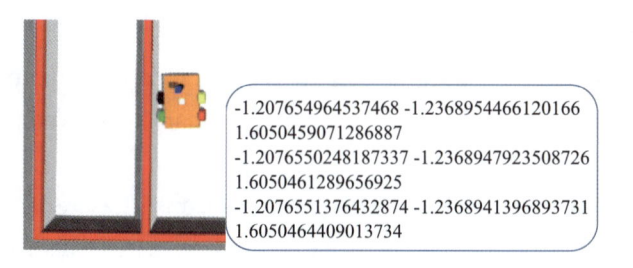

Figure 2-2-5　Position 2

Task 2　Learn Lidar Detection

The purpose of this task is to learn to use the lidar plug-in to detect the distance of obstacles around the maze robot.

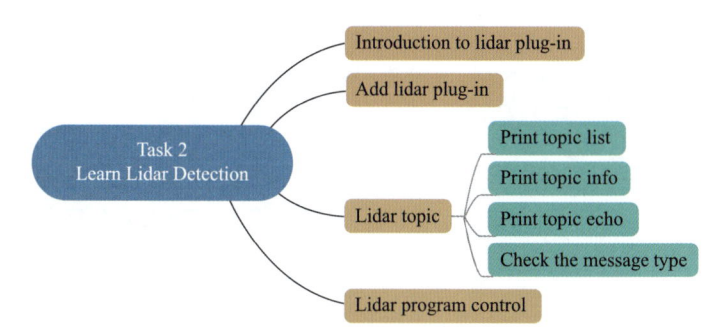

Ⅰ. Introduction to lidar plug-in

Ordinary single-line lidar generally has a transmitter and a receiver. The transmitter emits laser rays to the target in front, and the object will reflect the laser back, and then the lidar receiver can detect the reflected laser, as shown in Figure 2-2-6.

By calculating the time interval between transmission and feedback, and multiplying

Figure 2-2-6　Schematic diagram of laser detection

that by the speed of the laser, you can calculate how far the laser traveled(d).

The lidar plug-in is a plug-in in Gazebo used to detect the distance of obstacles. The dynamic link library libgazebo_ros_ray_sensor. so contains two light plug-ins, lidar and ultrasonic. You can receive commands to set the model's speed, and you can also provide feedback on the model's position and speed.

The prerequisite for the lidar plug-in to take effect is to correctly import the

plug-in in urdf and configure the relevant parameters.

By default, the lidar plug-in outputs the distance information detected by the current laser by publishing the topic /<pluginname>/out. You can choose to remap, such as /scan.

The message type of this topic is sensor_msgs/msg/LaserScan.

You can subscribe to this topic to get the current laser detected distance information.

Ⅱ. Add lidar plug-in

Add lidar code in urdf and configure related parameters.

```
        <gazebo reference="laser_link">   <!-- specify the effective
link-->
            <sensor name="laser_sensor" type="ray">  <!-- specify
name and type-->
                <pose>0 0 0.075 0 0 0</pose>  <!-- pose position, same
as joint-->
                <always_on>true</always_on>  <!-- always on-->
                <visualize>true</visualize>  <!-- is the blue light
visible-->
                <update_rate>60</update_rate>  <!-- update frequency-->
                <ray>
                    <scan>                    <!-- define the range of
scanning angle-->
                        <horizontal>    <!-- horizontal: horizontal
scan,vertical: vertical scan->
                            <samples>360</samples>   <!-- sampling
quantity,int-->
                            <resolution>1.000000</resolution>
<!-- resolution-->
                            <min_angle>0.000000</min_angle> < ! - -
scanning starting angle-->
                            <max_angle>6.280000</max_angle> < ! - -
scan termination angle-->
                        </horizontal>
                    </scan>
                    <range>                   <!-- define the scanning
distance range-->
                        <min>0.120000</min>   <!-- minimum distance-->
                        <max>3.5</max>         <!-- maximum distance-->
                        <resolution>0.015000</resolution><!-- resolution-->
                    </range>
                </ray>
                <plugin name="laserscan" filename="libgazebo_ros_
ray_sensor.so">   <!-- specify the plug-in name-->
                    <ros>
                        <remapping>~/out:=scan</remapping> < ! - -
remapping topic name-->
                    </ros>
```

Chapter2　Maze Robot Simulation Design | 131

```
                <output_type>sensor_msgs/LaserScan</output_
type>   <!-- output message type-->
                <frame_name>laser_link</frame_name>
            </plugin>
        </sensor>
    </gazebo>
```

Ⅲ. Lidar topic

After adding the lidar plug-in code, rebuild and launch tqd. launch. py.

```
$ cd ~/Maze robot _ws
$ colcon build
$ source install/setup. bash
$ ros2 launch tqd_Maze robot  tqd.launch.py
```

After starting, you can see that the blue area is the area scanned by the lidar, as shown in Figure 2-2-7.

1. Print topic list

```
$ ros2 topic list
```

The following information is output in the terminal:

```
/camera_image
/camera_info
/clock
/cmd_vel
/joint_states
/odom
/parameter_events
/rosout
/scan                    # lidar topic
/tf
```

Figure 2-2-7　Lidar scanning area

2. Print topic info

```
$ ros2 topic info /scan
```

The following information is output in the terminal:

```
Type: sensor_msgs/msg/LaserScan
Publisher count: 1
Subscription count: 0
```

There are no subscribers yet, so the Subscription count is 0.

3. Print topic echo

```
$ ros2 topic echo /scan
```

The following information is output in the terminal:

```
header:
  stamp:
    sec: 1280
    nanosec: 57000000
  frame_id: laser_link
angle_min: -1.5708999633789062
angle_max: 1.5708999633789062
angle_increment: 0.008751532062888145
time_increment: 0.0
scan_time: 0.0
range_min: 0.019999999552965164
range_max: 3.5
ranges:
- 0.09237094968557358
- 0.09239806979894638
- 0.09243228286504745
- 0.09247361123561859
- 0.09252206236124039
...
```

4. Check the message type

```
$ ros2 interface show sensor_msgs/msg/LaserScan
```

The following information is output in the terminal:

```
std_msgs/Header header # timestamp in the header is the acquisition
                               time of
        builtin_interfaces/Time stamp
                        int32 sec
                        uint32 nanosec
  string frame_id

float32 angle_min
float32 angle_max
float32 angle_increment
float32 time_increment
float32 scan_time
float32 range_min
float32 range_max
float32[] ranges
float32[] intensities
```

Chapter2   Maze Robot Simulation Design | 133

IV. Lidar program control

The following uses a Python program to interact with the lidar plug-in. Since the Python program is started manually externally, there is no need to create nodes in tqd_Maze robot , and the script file ray_sensor. py is created directly in the scripts directory.

Use the Python program to subscribe to/scan topic and obtain the obstacle distance around the maze robot in Gazebo.

```python
import rclpy
from rclpy.node import Node
from sensor_msgs.msg import LaserScan

def get_laser_callback(msg):
    region = msg. ranges
    # output horizontal right, right oblique 45, straight
    # ahead, left oblique 45, horizontal left distance
    print('right:', region[5])
    print('rf45 :', region[90])
    print('front:', region[180])
    print('lf45 :', region[270])
print('left :', region[355])

def get_laser():
    rclpy.init(args=None)
node_laser = Node('laser')
# create subscriber, topic/scan, callback get_laser_callback
        node_laser.create_subscription(LaserScan, '/scan', get_
laser_callback, 10)
    rclpy.spin(node_laser)
    rclpy.shutdown()

get_laser()
```

Video

Lidar detection

Run the program, move or rotate the maze robot, and observe the output data, as shown in Figure 2-2-8, Figure 2-2-9, and Figure 2-2-10.

right : 0.08618959784507751
rf45 : 0.12473881989717484
front : 2.763298988342285
lf45 : 0.11272557824850082
left : 0.08200465887784958
right : 0.08619077503681183
rf45 : 0.12473717331886292
front : 2.763324499130249
lf45 : 0.112727425992 48886

Figure 2-2-8   Position 1

right:0.27586621046066284
rf45:0.14005863666534424
front:1.7068243026733398
lf45:0.09896930307 149887
left:0.07306348532438278
right:0.27586814761161804
rf45:0.140061 57219409943
front:1.7067501 544952393
lf45:0.09896664321422577
left:0.07306203991174698
......

Figure 2-2-9　Position 2

right:0.09775447845458984
rf45:0.08569653332233429
front:0.06122323125600815
lf45:0.08788567781448364
left:0.25046828389167786
right:0.09775891155004501
rf45:0.08568809926509857
front:0.06121619418263434354
lf45:0.08787407726049423
left:0.250461 8465900421
......

Figure 2-2-10　Position 3

Task 3　Master the Maze Robot's Tracking Operation

The purpose of this task is to comprehensively use the two-wheel differential plug-in and the lidar plug-in to realize the maze robot tracking operation.

Task 3
Master the Maze Robot's
Tracking Operation

Principle of tracking operation

Tracking operation program control

Ⅰ. Principle of tracking operation

The best running trajectory of the maze robot is to walk along the center line of the maze, so that there is the same distance from the baffles on both sides to ensure that the maze robot can successfully reach the end of the maze, as shown in Figure 2-2-11.

Centerline　　　Left offset　　　Right offset　　　Forward

Figure 2-2-11　Location diagram

The lidar plug-in can detect the distance of the left, front and right baffles and determine the position of the maze robot in a cell.

The two-wheel differential plug-in can control the linear speed and

Chapter2   Maze Robot Simulation Design   135

angular speed of the maze robot's movement, straight line and turn, as shown in Figure 2-2-12.

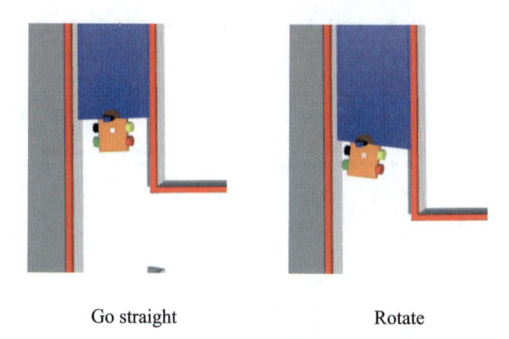

Go straight   Rotate

Figure 2-2-12   Action diagram

According to the baffle distance detected by the lidar plug-in, the target speed set by the two-wheel differential plug-in can be adjusted to realize the tracking operation of the maze robot.

II. Tracking operation program control

Create the script file track. py in the scripts directory.

Use the Python program to subscribe to /scan topic and obtain the obstacle distance around the maze robot in Gazebo.

Use a Python program to publish data to the /cmd_vel topic and set the target speed for the maze robot, as shown in Figure 2-2-13.

Video

Tracking operation

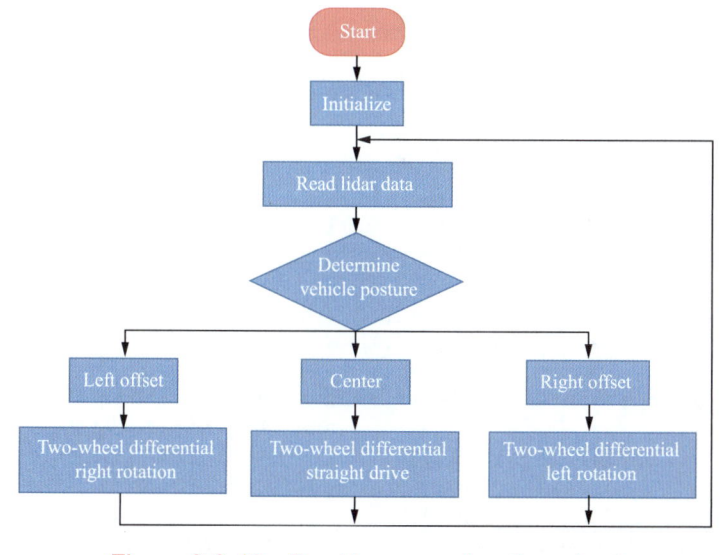

Figure 2-2-13   Tracking operation flow chart

It can be seen from Task 2 of this Project that when the maze robot is located on the center line of the maze, the distance between the horizontal left and right baffles is about 0. 085 meters, so 0. 085 is used as the threshold in the program to determine whether an offset has occurred.

```python
import rclpy
from rclpy. node import Node
from sensor_msgs. msg import LaserScan
from geometry_msgs. msg import Twist
import threading

class Track:
    def __init__(self) -> None:
        self.node = Node('mynode')
        self.msg = Twist()
        self.node.create_subscription(LaserScan, '/scan',
self. laser, 10)               # create a subscriber,/scan
        self.pub = self.node.create_publisher(Twist, '/cmd_
vel', 10)                      # create publisher, /cmd_vel
        self.rate = self.node.create_rate(50.0)
                               # create a Rate object with a fixed rate

        self.l_dis = 0          # assign  initial  values  to  each
attribute
        self.fl_dis = 0
        self.f_dis = 0
        self.fr_dis = 0
        self.r_dis = 0
        self.kp = 10            # correcting  the  strength  of  the
                               # vehicle posture
        self. speed_x = 0.1

    def laser(self, msg):       # /scan callback
        region = msg. ranges
        self.l_dis = region[355]
        self.fl_dis = region[270]
        self.f_dis = region[180]
        self.fr_dis = region[90]
        self.r_dis = region[5]

    def ros_spin(self):         # rclpy.spin(), call by child thread
        rclpy.spin(self.node)

    def posture_adjust(self):# determine whether it is offset
                               # and then correct it
        if self.l_dis<0.085:
            speed_z_temp = 0.5*(self.l_dis-0.085)   *self.kp
        elif self.r_dis<0.085:
            speed_z_temp = -0.5*(self.r_dis-0.085)   *self.kp
```

Chapter2　Maze Robot Simulation Design | 137

```python
        else:
            speed_z_temp = 0.0
        return speed_z_temp

    def move(self):                          # maze robot running
        while rclpy.ok():
            self.msg.linear.x = self.speed_x
            self.msg.angular.z = self.posture_adjust()
            self.pub.publish(self. msg)
            self.rate.sleep()                # implement fixed
                                             # rate publishing
            if 0. 05<self. f_dis<0. 08: # stop when the front fender
                                        # is detected
                self.msg.linear.x = 0.0
                self.msg.angular.z = 0.0
                self.pub.publish(self. msg)
                self.rate.sleep()
                break
def main():
    rclpy.init(args=None)
    track = Track()
    t = threading.Thread(None, target=track.ros_spin, daemon=True)
                         # call by child thread rclpy.spin()
    t.start()
    track.move()
if __name__ == '__main__':
    main()
```

Questions and Summary

1. How to coordinate the driving speed and actual speed of the control model?

2. How to adjust the angle range and distance range of lidar detection?

3. When using the horizontal sensor for tracking operation, we often find that it cannot be corrected in time. How can we improve it?

4. Multiple topics can be read and controlled in the program to achieve attitude detection and motion control of the model.

Project 3

Master the Intelligent Control of a Maze Robot

Learning objectives

1. Master how to calculate the coordinates of the maze robot in the maze.
2. Master how to process lidar detection data.
3. Master how to control the maze robot to turn accurately in the maze.

Task 1   Learn Coordinate Calculations

The purpose of this task is to calculate the coordinates of the maze robot in the maze based on its position in Gazebo.

I. Principle of coordinate calculation

The basic function of a maze robot is to run from the starting point to the end point. So how does the maze robot determine its position in the maze? The answer is based on the maze coordinates.

In the maze, each cell is represented by a coordinate. The starting point coordinate is defined as (0, 0), and the upper right corner is defined as (F, F), as shown in Figure 2-3-1.

There are two ways to calculate the coordinates of the maze robot in the maze:

Method 1: Determine which cell it belongs to according to its position in Gazebo, thereby determining the maze coordinates.

Chapter2　Maze Robot Simulation Design　**139**

Method 2: Calculate the maze coordinates based on the direction and distance of movement in Gazebo.

The method 1 is to calculate coordinates by using absolute positions. However, since Gazebo is a simulation environment, the model will move over time, so the absolute position of the maze robot in Gazebo is not accurate. Therefore, method 2 is usually used to calculate the maze coordinates.

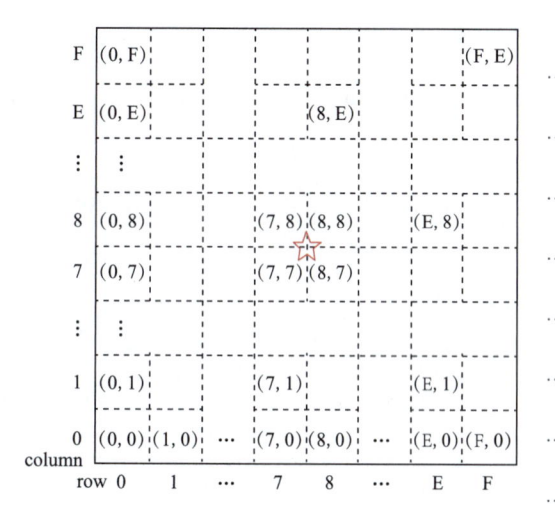

Figure 2-3-1　Maze coordinate diagram

The two-wheel differential plug-in can obtain the position of the maze robot in Gazebo and drive the maze robot to run, which can be used to calculate the maze coordinates.

① Record the position of the maze robot in Gazebo before driving it.

② Drive the maze robot to start running.

③ Whenever the maze robot moves more than one cell, the coordinate is +1 in the direction of movement.

After determining the starting coordinates of the maze robot, the new coordinates can be calculated by looping the above three steps, as shown in Figure 2-3-2.

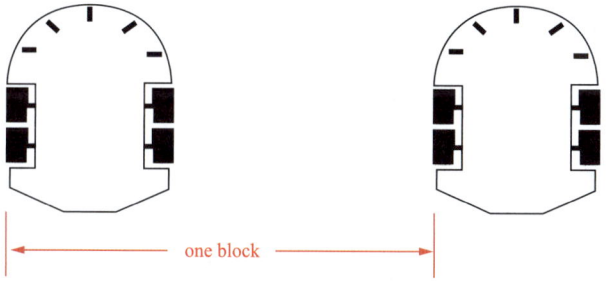

Figure 2-3-2　Schematic diagram of coordinate update

II. Coordinate calculation program control

1. Start tqd_Maze robot

First start tqd. launch in the tqd_Maze robot function package. py

```
$ cd ~/Maze robot _ws
$ source install/setup. bash
$ ros2 launch tqd_Maze robot  tqd.launch.py
```

2. Create coordinate. py

Create the script file coordinate. py in the scripts directory. Write the following code:

```
import rclpy
from rclpy.node import Node
from std_msgs.msg import String
from geometry_msgs.msg import Twist
from nav_msgs.msg import Odometry
from sensor_msgs.msg import LaserScan
import tf_transformations
import threading

class Drive(object):
    def __init__(self) -> None:
        self.valueinit()
        self.ros_register()

    def valueinit(self):
        self.l_dis = 0
        self.fl_dis = 0
        self.f_dis = 0
        self.fr_dis = 0
        self.r_dis = 0
        self.position_x = 0
        self.position_y = 0
        self.yaw = 1.5708

        self.blocksize = 0.178                    # maze cell default size
        self.kp = 10

        self.speed_x = 0.10
        self.speed_z = 0.35

    def ros_register(self):
        self.msg = Twist()
        self.string = String()
        self.node = Node('mynode')
        self.node.create_subscription(LaserScan, '/scan',
self.laser, 10)                   # create a subscriber,/scan
        self.node.create_subscription(Odometry, '/odom', self.
odom, 10)                         # create a subscriber,/odom
        self.pub = self.node.create_publisher(Twist, '/cmd_
vel', 10)                                  # create publisher,/cmd_vel
        self.rate = self.node.create_rate(50.0)
```

Chapter2　Maze Robot Simulation Design　141

```python
    def ros_spin(self):          # rclpy.spin(),call by child thread
        rclpy.spin(self.node)

    def laser(self, msg):              # /scan callback
        region = msg. ranges
        self.l_dis = region[340]
        self.fl_dis = region[270]
        self.f_dis = region[180]
        self.fr_dis = region[90]
        self.r_dis = region[20]

    def get_laser(self):
        return self.l_dis, self.fl_dis, self.f_dis, self.fr_
dis, self.r_dis

    def odom(self, msg):              # /odom callback
        position = msg.pose.pose.position
        self.position_x = position.x
        self.position_y = position.y
        quaternion = (
            msg.pose.pose.orientation.x,
            msg.pose.pose.orientation.y,
            msg.pose.pose.orientation.z,
            msg.pose.pose.orientation.w)
        euler = tf_transformations.euler_from_quaternion(quaternion)
        self.yaw = euler[2]

    def get_odom(self):
        return self.position_x, self.position_y, self.yaw

    def posture_adjust(self): # determine whether it is offset
                             # and then correct it
        if self.f_dis>0.09:
            if self.fl_dis<0.118:
                speed_z_temp = 0.5*(self. fl_dis-0.118) *self.kp
            elif self.fr_dis<0.118:
                speed_z_temp = -0.5*(self. fr_dis-0.118) *self.kp
            else:
                speed_z_temp = 0.0
        else:
            if self.l_dis<0.085:
                speed_z_temp = 0.5*(self.l_dis-0.085) *self.kp
            elif self.r_dis<0.085:
                speed_z_temp = -0.5*(self.r_dis-0.085) *self.kp
            else:
                speed_z_temp = 0.0
        return speed_z_temp

    def move(self, numblock=1):      # driving the maze robot to run
        flag = 1
        while rclpy.ok():
```

```
                        self.msg.linear.x = self.speed_x
                        self.msg.angular.z = self.posture_adjust()
                        self.pub.publish(self.msg)
                        self.rate.sleep()
                        if flag: # when the drive is first initiated, self.get_
                                # odom() returns 0, hence this condition is added
                            tempx, tempy, tempz = self.get_odom()
                            if (not tempx) and (not tempy):
                                continue
                            flag = 0
                        if abs(self.get_odom()[0] - tempx)>=self.blocksize*
numblock or abs(self.get_odom()[1] - tempy)>=self.blocksize*numblock:
                    # end the loop when the moving distance exceeds the set value
                            self.msg.linear.x = 0.0
                            self.msg.angular.z = 0.0
                            self.pub.publish(self.msg)
                            self.rate.sleep()
                            break


    def main():
        rclpy.init(args=None)
        drive = Drive()
        t = threading.Thread(None, target=drive.ros_spin, daemon=True)
        t.start()
        drive.move(10)                          # move 10 cells

    if __name__ == '__main__':
        try:
            main()
        finally:
            rclpy.shutdown()
```

Modify the number of cells walked by drive.move(10) and observe the number of cells actually running in the maze. If they do not match, as shown in Figure 2-3-3, appropriately modify the size of self. blocksize until the two match, as shown in Figure 2-3-4.

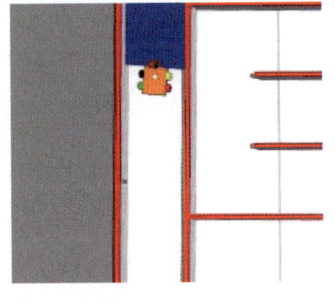

Figure 2-3-3   Does not match

Figure 2-3-4   Matches

Chapter2　Maze Robot Simulation Design　143

Task 2　Learn Intersection Detection

The purpose of this task is to determine the wall data at the current location based on the lidar detection data, that is, whether the left, front, and right sides are walls or intersections respectively.

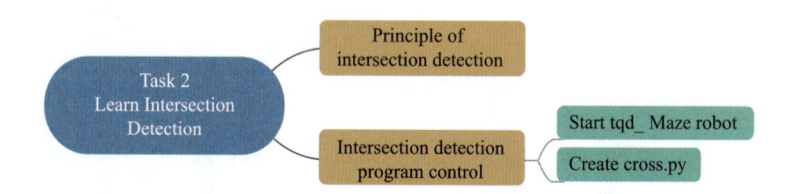

I . Principle of intersection detection

The maze robot uses lidar to detect the distance of the surrounding baffles. The detection distance can be specified in the urdf model. In order to improve detection accuracy and facilitate expansion, the lower detection limit is usually selected around 0.02 m. If it is too large or too small, it may cause false detection; The detection upper limit is usually chosen to be a value greater than 1 m. If it is too small, it will not be conducive to later program optimization.

When the maze robot walks in the cell, the most perfect running trajectory is to walk along the center line. If offset occurs, the most extreme case is that it is close to the left baffle or close to the right baffle, as shown in Figure 2-3-5.

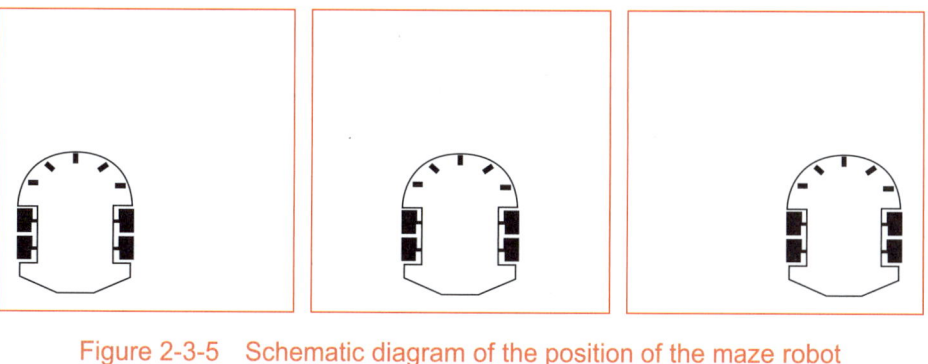

Figure 2-3-5　Schematic diagram of the position of the maze robot

When the maze robot is in the middle of the cell, the threshold distance from the baffles on both sides is approximately 0.08 m; when the maze robot is close to the left or right side, the distance from the baffles on the other side is approximately 0.10 m, as shown in Figure 2-3-6.

Therefore, when the distance between the left, front and right walls detected

by the maze robot is greater than 0.12, it can be concluded that there is an intersection in the cell in that direction, as shown in Figure 2-3-7.

Combined with the coordinates of the maze robot, the wall data of all passing coordinates can be saved.

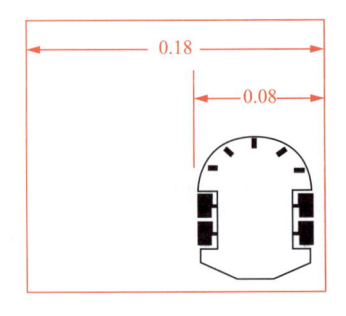

Figure 2-3-6　Schematic diagram of cell and robot dimensions

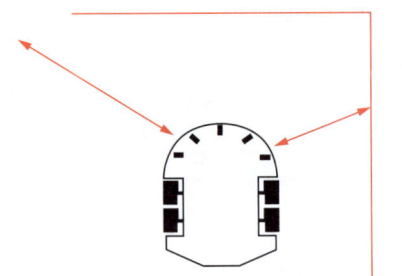

Figure 2-3-7　Schematic diagram of intersection

Ⅱ. Intersection detection program control

1. Start tqd_Maze robot

First start tqd. launch. py in the tqd_Maze robot function package.

```
$ cd ~/Maze robot _ws
$ source install/setup.bash
$ ros2 launch tqd_Maze robot  tqd.launch.py
```

2. Create cross. py

Create the script file cross. py in the scripts directory. Copy the code in Task 1 of this project and modify the main program as follows:

```
def main()   :
    rclpy. init(args=None)
    drive = Drive()
    t = threading. Thread(None, target=drive. ros_spin, daemon=True)
    t. start()
    x, y = 0, 0
    while 1:
        drive. move()
        y = y+1
        l_dis, fl_dis, f_dis, fr_dis, r_dis = drive. get_laser()
        leftpath = '1' if l_dis>0. 12 else '0'
        backpath = '1'
        rightpath = '1' if r_dis>0. 12 else '0'
        frontpath = '1' if f_dis>0. 12 else '0'
            print(f"({x}, {y}), {leftpath} {backpath} {rightpath} {frontpath}")
```

● Video

Intersection detection

Chapter2　Maze Robot Simulation Design | 145

```
    if y==10:
        break
```

Run the program and observe the output data, as shown in Figure 2-3-8.

You can also try to comment out the following statements, run the program, move the maze robot, and observe the output data, as shown in Figure 2-3-9 and Figure 2-3-10.

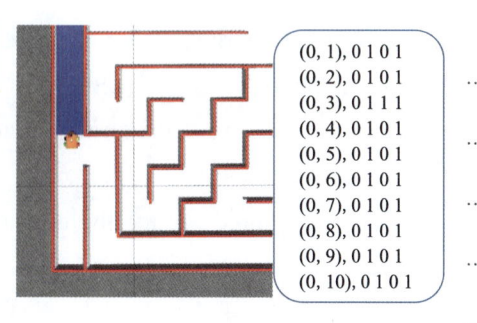

Figure 2-3-8　Determine cell wall information in real time

```
# drive. move()
# y = y+1
```

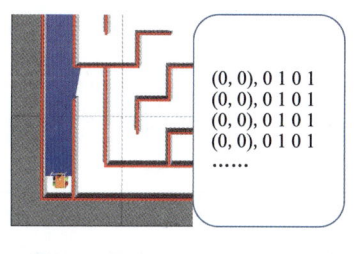

Figure 2-3-9　Position 1

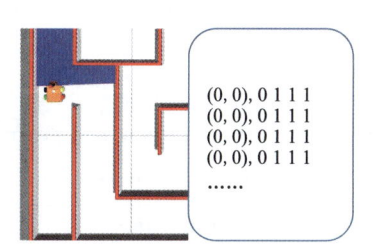

Figure 2-3-10　Position 2

Task 3　Master Turning Control

The purpose of this task is to control the maze robot to make precise turns based on changes in yaw angle.

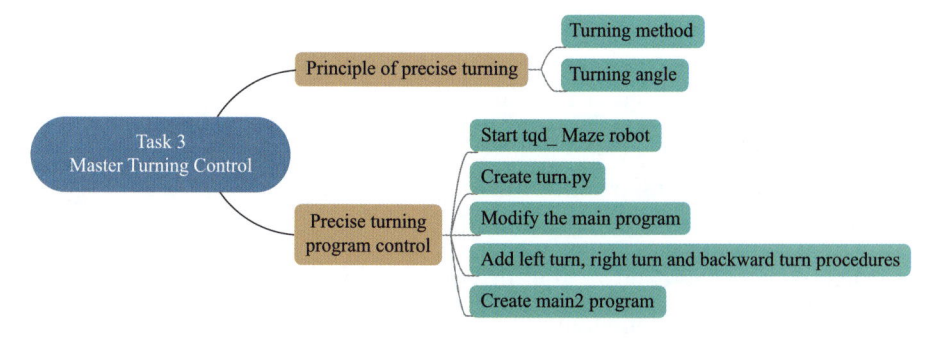

Ⅰ. Principle of precise turning

When the maze robot runs in the maze, in addition to running straight, it will

also encounter a large number of turns. Whether the turning is accurate or not has a huge impact on the subsequent operation of the maze robot. Accurate turns need to consider two factors, the turning style and the turning angle.

1. Turning method

There are three commonly used turning methods for maze robots, as shown in Figure 2-3-11.

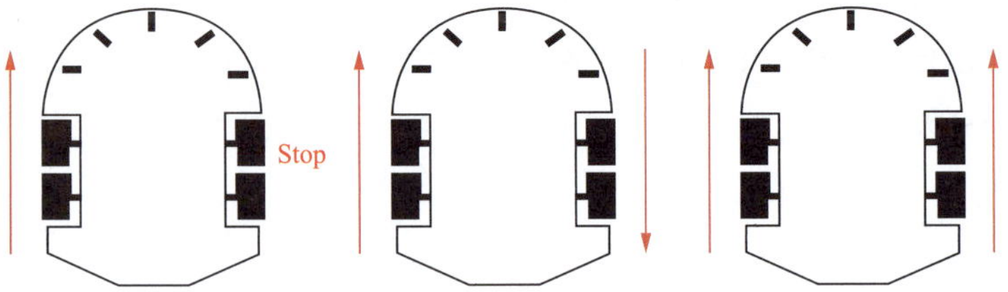

Figure 2-3-11　Three turning methods

Method 1: The inner wheel stops and the outer wheel rotates forward. This turning method is equivalent to turning 1/4 arc with the inner wheel as the center of the circle, and the turning radius is larger.

Method 2: The inner wheel rotates reversely and the outer wheel rotates forward, and the speed of the two wheels is the same. This turning method is equivalent to turning 1/4 arc with the center of the maze robot as the center of the circle, with the smallest turning radius and the highest stability.

Method 3: Both the inside wheel and the outside wheel rotate forward, but the speed of the outside wheel is higher. This turning method is the fastest, but it is slightly more difficult to control.

Next, we use method 2 as the turning method of the maze robot.

2. Turning angle

The size of the turning angle mainly depends on when to stop turning. As shown in Figure 2-3-12, the maze robot's turning angle control mainly uses two methods:

Method 1: Delay control, after driving the maze robot to enter the rotation state, delay for a certain period of time, and then end the rotation. This control method is the simplest, but has poorer accuracy.

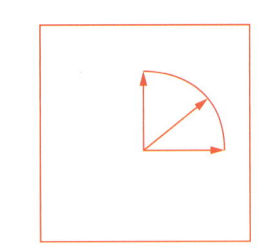

Figure 2-3-12　Diagram of turning angle

Method 2: Sensor detection control, by using sensors to detect the attitude

Chapter2 Maze Robot Simulation Design | 147

changes of the maze robot, and when the preset value is reached, the rotation ends. This control method relies on sensor detection and has higher accuracy.

In Gazebo simulation, there are a variety of sensors that can be used to detect attitude changes of the maze robot, such as the yaw angle provided by the two-wheel differential plug-in, as shown in Figure 2-3-13.

When the maze robot's head turns to the left, there is a jump from π to -π. In order to facilitate calculation, the yaw angle is corrected and all yaw angle data is increased by π, as shown in Figure 2-3-14. In the program, π is calculated according to 3. 141 59.

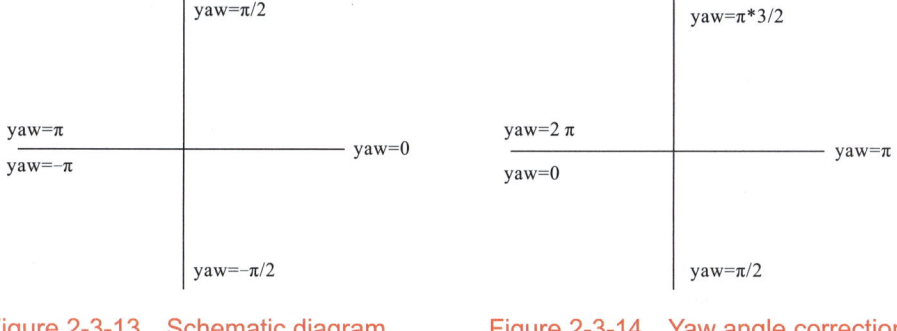

Figure 2-3-13　Schematic diagram of yaw angle

Figure 2-3-14　Yaw angle correction

Next we will choose the yaw angle as the control method of the turning angle.

II . Precise turning program control

1. Start tqd_Maze robot

First start tqd. launch. py in the tqd_Maze robot function package.

```
$ cd ~/Maze robot _ws
$ source install/setup.bash
$ ros2 launch tqd_Maze robot  tqd.launch.py
```

2. Create turn. py

Create the script file turn. py in the scripts directory. Copy the code in Task 1 of this project and add the assignment of the yaw angle in the valueinit() method:

```
def valueinit(self)   :
    self. l_dis = 0
    self. fl_dis = 0
    self. f_dis = 0
    self. fr_dis = 0
```

Video

Turn control

```
        self.r_dis = 0
        self.position_x = 0
        self.position_y = 0
        self.yaw = 0
        self.fyaw = 4.71        # yaw angle when the front of the
                                # car is moving forward
        self.ryaw = 3.14        # yaw angle when the front of the
                                # car is turning to the right
        self.byaw = 1.57        # yaw angle when the front of the
                                # car is down
        self.lyaw0 = 0     # yaw angle 0 when the front of the car
                           # is turning to the left
        self.lyaw1 = 6.28  # yaw angle 1 when the front of the
                           # car is turning to the left

        self.blocksize = 0.175
        self.kp = 10

        self.speed_x = 0.10
        self.speed_z = 0.35
```

3. Modify the main program

```
def main():
    rclpy.init(args=None)
    drive = Drive()
    t = threading.Thread(None, target=drive.ros_spin, daemon=True)
    t.start()
    x, y = 0, 0
    while 1:
        drive. move()
        y = y+1
        l_dis, fl_dis, f_dis, fr_dis, r_dis = drive.get_laser()
        leftpath = '1' if l_dis>0.12 else '0'
        backpath = '1'
        rightpath = '1' if r_dis>0.12 else '0'
        frontpath = '1' if f_dis>0.12 else '0'

        if rightpath=='1':
            drive.turn(0)
            break

        if y==10:
            break
```

Run the program. When the maze robot detects an intersection on the right, it turns right and then stops.

4. Add left turn, right turn and backward turn procedures

When the maze robot searches or sprints, the actions involved are left turn, right turn and backward turn. Therefore, it is also necessary to deeply determine

which direction the current yaw is and which direction it needs to turn to, and add
right turn, left turn and backward turn code.

```python
def turnright(self):
    flag = 1
    while rclpy.ok():
        self.msg.linear.x = 0.0
        self.msg.angular.z = -self.speed_z
        self.pub.publish(self.msg)
        self.rate.sleep()
        if flag:
            oldyaw = self.get_odom()[2]
            if not oldyaw:
                continue
            flag = 0
        # Considering the mutation of 0 and 6.28, so
           dividing the whole process into four parts.

        if 4.41<oldyaw<5.01:
            if self.ryaw-0.1<self.yaw<self.ryaw+0.1:
                self.msg.linear.x = 0.0
                self.msg.angular.z = 0.0
                self.pub.publish(self.msg)
                # rclpy.spin_once(self.node)
                self.rate.sleep()
                break

        elif 2.84<oldyaw<3.44:
            if self.byaw-0.1<self.yaw<self.byaw+0.1:
                self.msg.linear.x = 0.0
                self.msg.angular.z = 0.0
                self.pub.publish(self.msg)
                # rclpy.spin_once(self.node)
                self.rate.sleep()
                break

        elif 1.27<oldyaw<1.87:
            if self.yaw<self.lyaw0+0.1:
                self.msg.linear.x = 0.0
                self.msg.angular.z = 0.0
                self.pub.publish(self.msg)
                # rclpy.spin_once(self.node)
                self.rate.sleep()
                break

        elif oldyaw<0.3 or oldyaw>5.98:
            if self.fyaw-0.1<self.yaw<self.fyaw+0.1:
                self.msg.linear.x = 0.0
                self.msg.angular.z = 0.0
                self.pub.publish(self.msg)
                # rclpy.spin_once(self.node)
```

```python
                    self.rate.sleep()
                    break

    def turnleft(self):
        flag = 1
        while rclpy.ok():
            self.msg.linear.x = 0.0
            self.msg.angular.z = self.speed_z
            self.pub.publish(self.msg)
            self.rate.sleep()
            if flag:
                oldyaw = self.yaw
                if not oldyaw:
                    continue
                flag = 0
            if 4.41<oldyaw<5.01:
                if self.yaw>self.lyaw1-0.1:
                    self.msg.linear.x = 0.0
                    self.msg.angular.z = 0.0
                    self.pub.publish(self.msg)
                    # rclpy.spin_once(self.node)
                    self.rate.sleep()
                    break
            elif 2.84<oldyaw<3.44:
                if self.fyaw-0.1<self.yaw<self.fyaw+0.1:
                    self.msg.linear.x = 0.0
                    self.msg.angular.z = 0.0
                    self.pub.publish(self.msg)
                    # rclpy.spin_once(self.node)
                    self.rate.sleep()
                    break
            elif 1.27<oldyaw<1.87:
                if self.ryaw-0.1<self.yaw<self.ryaw+0.1:
                    self.msg.linear.x = 0.0
                    self.msg.angular.z = 0.0
                    self.pub.publish(self.msg)
                    # rclpy.spin_once(self.node)
                    self.rate.sleep()
                    break
            elif oldyaw<0.3 or oldyaw>5.98:
                if self.byaw-0.1<self.yaw<self.byaw+0.1:
                    self.msg.linear.x = 0.0
                    self.msg.angular.z = 0.0
                    self.pub.publish(self.msg)
                    # rclpy.spin_once(self.node)
                    self.rate.sleep()
                    break

    def turnback(self):
        flag = 1
        while rclpy.ok():
```

Chapter2   Maze Robot Simulation Design   151

```
            self.msg.linear.x = 0.0
            self.msg.angular.z = -self.speed_z
            self.pub.publish(self.msg)
            self.rate.sleep()
            if flag:
                oldyaw = self.get_odom()    [2]
                if not oldyaw:
                    continue
                flag = 0
            if 4.41<oldyaw<5.01:
                if self.byaw-0.1<self.yaw<self.byaw+0.1:
                    self.msg.linear.x = 0.0
                    self.msg.angular.z = 0.0
                    self.pub.publish(self.msg)
                    # rclpy.spin_once(self.node)
                    self.rate.sleep()
                    break
            elif 2.84<oldyaw<3.44:
                if self.yaw<self.lyaw0+0.1:
                    self.msg.linear.x = 0.0
                    self.msg.angular.z = 0.0
                    self.pub.publish(self.msg)
                    # rclpy.spin_once(self.node)
                    self.rate.sleep()
                    break
            elif 1.27<oldyaw<1.87:
                if self.fyaw-0.1<self.yaw<self.fyaw+0.1:
                    self.msg.linear.x = 0.0
                    self.msg.angular.z = 0.0
                    self.pub.publish(self.msg)
                    # rclpy.spin_once(self.node)
                    self.rate.sleep()
                    break
            elif oldyaw<0.3 or oldyaw>5.98:
                if self.ryaw-0.1<self.yaw<self.ryaw+0.1:
                    self.msg.linear.x = 0.0
                    self.msg.angular.z = 0.0
                    self.pub.publish(self.msg)
                    # rclpy.spin_once(self.node)
                    self.rate.sleep()
                    break
```

5. Create main2 program

```
def main2():
    rclpy.init(args=None)
    drive = Drive()
    t = threading.Thread(None, target=drive.ros_spin, daemon=True)
    t.start()
    x, y = 0, 0
    while 1:
```

```
drive.move()
y = y+1
l_dis, fl_dis, f_dis, fr_dis, r_dis = drive.get_laser()
leftpath = '1' if l_dis>0.12 else '0'
backpath = '1'
rightpath = '1' if r_dis>0.12 else '0'
frontpath = '1' if f_dis>0.12 else '0'

if rightpath=='1':
    drive.turnright()
    break

if leftpath=='1':
    drive.turnleft()
    break

if frontpath=='0':
    drive.turnback()
    break

if y==10:
    break
```

Execute the main2 program and observe the operation of the maze robot. You can also try to move the maze robot to other locations, such as an area with an intersection on the left or a barrier in front, and execute the program again to observe the phenomenon.

Questions and Summary

1. How to improve the detection accuracy of walls?

2. What are the methods of coordinate calculation and what are their advantages and disadvantages?

3. There are many ways to turn. Turning while traveling is faster, but the movement process is more complicated; turning on the spot is slower, but the control is simpler.

Chapter **3**

Maze Robot Simulation Practice

After mastering the basic control methods of maze robot simulation, you can try to compete in the maze robot simulation competition. In this chapter, it mainly explains how to implement various program logic of the maze robot in the simulation environment, such as search strategies, intelligent algorithms for information access, etc.; finally, use the written program to conduct a simulation competition.

Chapter3　Maze Robot Simulation Practice　155

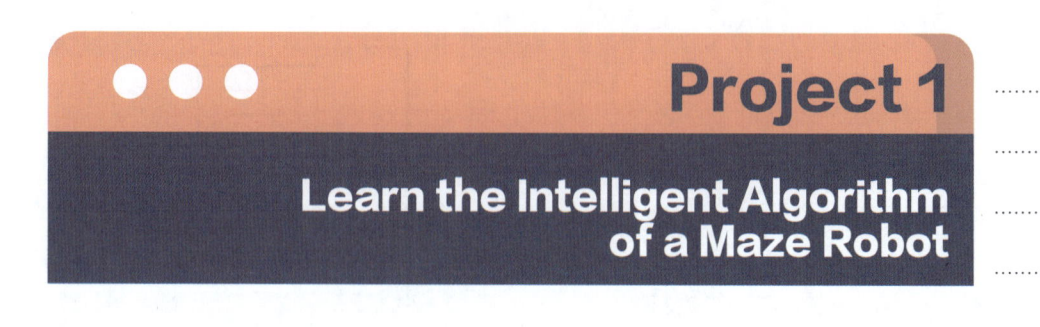

Project 1

Learn the Intelligent Algorithm of a Maze Robot

Learning objectives

1. Understand maze search strategies and their advantages and disadvantages.

2. Understand how to store and read maze information.

3. Understand how to plan the optimal path in the maze.

Task 1　Understand the Maze Search Strategy

The purpose of this task is to understand the way the maze robot searches the maze and master the commonly used search rules.

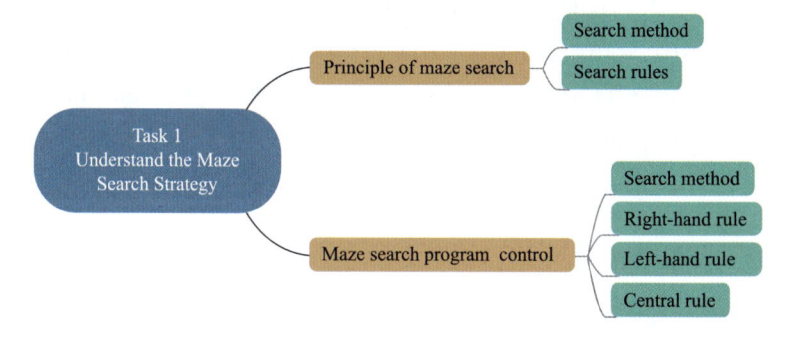

Ⅰ. Principle of maze search

On the first run, the maze is unknown to the maze robot and it is uncertain how to walk to get to the end. So you need to search the maze until you reach the end.

1. Search method

There are usually two ways for maze robots to search mazes:

Method 1: Reach the end point as soon as possible, as shown in Figure 3-1-1.

Method 2: Search the entire maze, as shown in Figure 3-1-2.

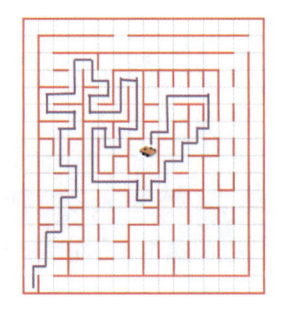

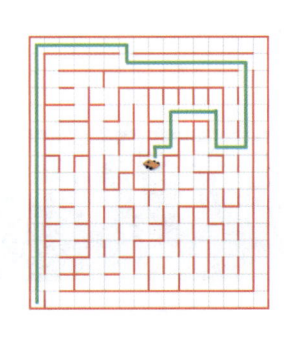

Figure 3-1-1   Partial search   Figure 3-1-2   Full maze search

There are pros and cons to both approaches. Although using the first method can shorten the time required to search the maze, it may not necessarily be able to obtain information about the entire maze map. If the path found is not the optimal path in the maze, this will affect the final sprint time of the maze robot. Using the second method, you can get information about the entire maze map, so you can find the optimal path. However, the search time used by this method will be longer.

For this task, choose method 1 to search.

2. Search rules

When the maze robot searches the maze, it needs to record its own maze coordinates and corresponding wall information at all times. Aimless search will greatly affect the efficiency. In order to save search time, some search rules will be used to improve search efficiency when searching.

There are three commonly used basic search rules: right-hand rule, left-hand rule and central rule, as shown in Figure 3-1-3.

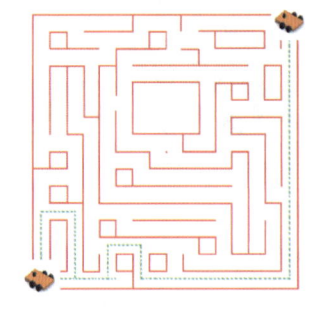

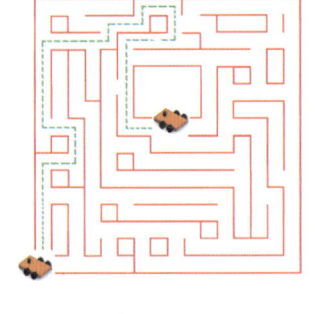

（a）Right-hand rule   （b）Left-hand rule   （c）Central rule

Figure 3-1-3   Right-hand rule, left-hand rule, and central rule

Right-hand rule: When there are multiple directions for the maze robot to choose from, turn right first, then go straight, and finally turn left.

Chapter3　Maze Robot Simulation Practice | 157

Left-hand rule: When there are multiple directions for the maze robot to choose from, turn left first, then go straight, and finally turn right.

Central rule: When there are multiple directions for the maze robot to choose from, it will give priority to turning in the direction of the end point.

II. Maze search program control

Next, implement the search and rules of the maze robot.

First start tqd. launch. py in the tqd_Maze robot function package.

Video

Maze search

```
$ cd ~/Maze robot _ws
$ source install/setup. bash
$ ros2 launch tqd_Maze robot  tqd. launch. py
```

Copy the complete program file contest. py.

In order to separate the underlying driver and top-level logic, create class Maze robot and inherit class Drive. class Drive is used as the driver class, and class Maze robot is used as the logic class. All the upper-level logic of the maze robot is implemented in class Maze robot.

The following mainly introduces methods related to search strategies. Please scan the QR code to view the complete code.

1. Search method

```
    def mazesearch(self):
        while 1:
            if self. destinationcheck():
    # if you reach the end point, return to the starting point and
    # search partially

                destination = (self. mouse_x, self. mouse_y)
                self. turnback()
                self. objectgoto(0, 0)
                return (1, destination)
            else:                        #otherwise, keep searching
                crosswaycount = self. crosswaycheck()
                if crosswaycount:
                    if crosswaycount>1:
                        self. crosswaystack. append((self. mouse_x,
self. mouse_y))
                        self. crosswaychoice()
                        self. moveoneblock()
                    if crosswaycount==1:
                        self. crosswaychoice()
                        self. moveoneblock()
                else:
                    self. turnback()
```

File

Complete code

```
                                        self. objectgoto(*self. crosswaystack. pop() )
```

2. Right-hand rule

```python
def rightmethod(self):
    frontpath = '1' if self. f_dis > 0. 2 else '0'
    rightpath = '1' if self. r_dis > 0. 2 else '0'
    backpath = '1'
    leftpath = '1' if self. l_dis > 0. 2 else '0'
    if rightpath == '1': # there is a road on the right side
        if self. dir == 0:
            flag = self. mapblock[self. mouse_x+1, self. mouse_y][0]
        elif self. dir == 1:
            flag = self. mapblock[self. mouse_x, self. mouse_y-1][0]
        elif self. dir == 2:
            flag = self. mapblock[self. mouse_x-1, self. mouse_y][0]
        elif self. dir == 3:
            flag = self. mapblock[self. mouse_x, self. mouse_y+1][0]
        if flag == '0':
            self. turnright()
            return

    if frontpath == '1':          # there is a road ahead
        if self. dir == 0:
            flag = self. mapblock[self. mouse_x, self. mouse_y+1][0]
        elif self. dir == 1:
            flag = self. mapblock[self. mouse_x+1, self. mouse_y][0]
        elif self. dir == 2:
            flag = self. mapblock[self. mouse_x, self. mouse_y-1][0]
        elif self. dir == 3:
            flag = self. mapblock[self. mouse_x-1, self. mouse_y][0]
        if flag == '0':
            return

    if leftpath == '1':  # there is a road on the left side
        if self. dir == 0:
            flag = self. mapblock[self. mouse_x-1, self. mouse_y][0]
        elif self. dir == 1:
            flag = self. mapblock[self. mouse_x, self. mouse_y+1][0]
        elif self. dir == 2:
            flag = self. mapblock[self. mouse_x+1, self. mouse_y][0]
        elif self. dir == 3:
            flag = self. mapblock[self. mouse_x, self. mouse_y-1][0]
        if flag == '0':
            self. turnleft()
            return
```

3. Left-hand rule

```python
def leftmethod(self):
    frontpath = '1' if self. f_dis > 0. 2 else '0'
    rightpath = '1' if self. r_dis > 0. 2 else '0'
```

```
        backpath = '1'
        leftpath = '1' if self. l_dis > 0. 2 else '0'
        if leftpath == '1':   # there is a road on the left side
            if self. dir == 0:
                flag = self. mapblock[self. mouse_x-1, self. mouse_y][0]
            elif self. dir == 1:
                flag = self. mapblock[self. mouse_x, self. mouse_y+1][0]
            elif self. dir == 2:
                flag = self. mapblock[self. mouse_x+1, self. mouse_y][0]
            elif self. dir == 3:
                flag = self. mapblock[self. mouse_x, self. mouse_y-1][0]
            if flag == '0':
                self. turnleft()
                return

        if frontpath == '1': # there is a road ahead
            if self. dir == 0:
                flag = self. mapblock[self. mouse_x, self. mouse_y+1][0]
            elif self. dir == 1:
                flag = self. mapblock[self. mouse_x+1, self. mouse_y][0]
            elif self. dir == 2:
                flag = self. mapblock[self. mouse_x, self. mouse_y-1][0]
            elif self. dir == 3:
                flag = self. mapblock[self. mouse_x-1, self. mouse_y][0]
            if flag == '0':
                return

        if rightpath == '1': # there is a road on the right side
            if self. dir == 0:
                flag = self. mapblock[self. mouse_x+1, self. mouse_y][0]
            elif self. dir == 1:
                flag = self. mapblock[self. mouse_x, self. mouse_y-1][0]
            elif self. dir == 2:
                flag = self. mapblock[self. mouse_x-1, self. mouse_y][0]
            elif self. dir == 3:
                flag = self. mapblock[self. mouse_x, self. mouse_y+1][0]
            if flag == '0':
                self. turnright()
                return
```

4. Central rule

```
def centralmethod(self):
    if self. mouse_x < 8:
        if self. mouse_y < 8:    # Maze robot is located in
                                 the bottom left corner
            if self. dir == 0:
                self. frontrightmethod()
            elif self. dir == 1:
                self. frontleftmethod()
            elif self. dir == 2:
                self. leftmethod()
```

```
                elif self. dir == 3:
                    self. rightmethod()
            else:                          # Maze robot is located in
                                               the top left corner
                if self. dir == 0:
                    self. rightmethod()
                elif self. dir == 1:
                    self. frontrightmethod()
                elif self. dir == 2:
                    self. frontleftmethod()
                elif self. dir == 3:
                    self. leftmethod()
        else:
            if self. mouse_y < 8:    # Maze robot is located in
                                          the bottom right corner
                if self. dir == 0:
                    self. frontleftmethod()
                elif self. dir == 1:
                    self. leftmethod()
                elif self. dir == 2:
                    self. rightmethod()
                elif self. dir == 3:
                    self. frontrightmethod()
            else:                          # Maze robot is located in
                                               the top right corner
                if self. dir == 0:
                    self. leftmethod()
                elif self. dir == 1:
                    self. rightmethod()
                elif self. dir == 2:
                    self. frontrightmethod()
                elif self. dir == 3:
                    self. frontleftmethod()
```

Task 2   Master the Storage and Reading of Maze Information

The purpose of this task is to master the storage and reading of maze information.

Chapter3　Maze Robot Simulation Practice | 161

Ⅰ. Principle of maze information access

The purpose of the maze robot searching the maze is to find the end point. During the search process, the maze map information needs to be recorded, so that the saved information can be directly read during the second walk and sprint to the end point.

The maze is composed of 16×16 cells. It is obvious to create a two-dimensional array, x and y correspond to the row and col of the array respectively, and the cell wall data is saved as the corresponding value. Every time the maze robot updates its coordinates, it should save the wall data until it reaches the end point.

1. Storage method

For the wall data of a cell, the direction in which the maze robot comes in must be the intersection (except the starting point). The other three sides need to be judged based on the data detected by the lidar plug-in. If it is greater than 0.12 m, it is an intersection, and if it is less than 0.12 m, it is a baffle. The presence or absence of an intersection corresponds to 1 or 0, so this task uses a string of length 4, with each bit using 1 or 0, to represent the wall data of a cell, such as "0001", as shown in Figure 3-1-4.

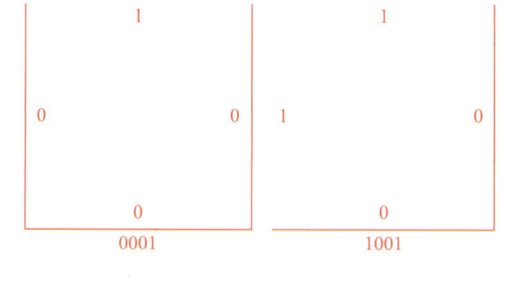

Figure 3-1-4　Schematic diagram of wall data

① The first digit represents the left wall of the cell, 1 represents an intersection, and 0 represents no intersection;

② The second digit represents the wall below the cell, 1 represents an intersection, 0 represents no intersection;

③ The third digit represents the wall on the right side of the cell, 1 represents an intersection, and 0 represents no intersection;

④ The fourth digit represents the wall above the cell, 1 represents an intersection, 0 represents no intersection;

Wall information storage is shown in Figure 3-1-5.

2. Direction change

Because when the maze robot detects the surrounding walls, it uses the maze robot itself as a reference, which is the relative direction data. When saving data into an array, it is recommended to convert it into absolute direction data, as

shown in Figure 3-1-6.

Relative direction: left, rear, right, front.

Absolute direction: left bottom right up.

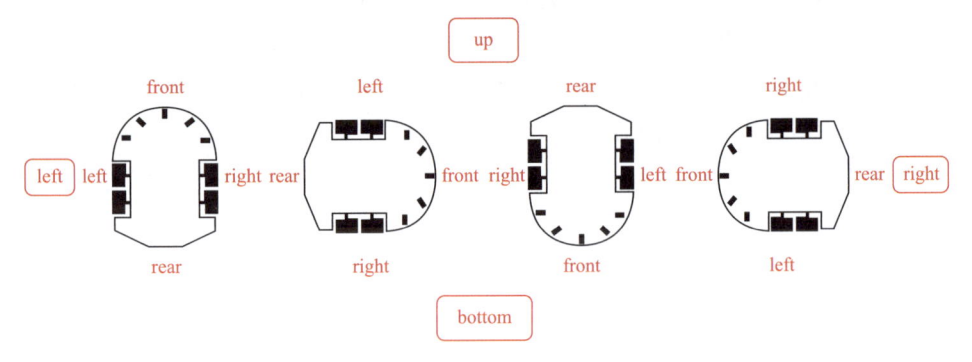

0110	1010	1100	0000	0110	1010	1000	0111
0011	1100	0011	1010	1011	1010	1000	0101
0110	1011	1010	1100	0010	1110	1010	1001
0101	0010	1010	1011	1000	0011	1010	1100
0111	1010	1010	1010	1010	1010	1100	0101
0101	0110	1100	0110	1100	0000	0101	0101
0101	0101	0011	1101	0011	1010	1011	1101
0001	0011	1000	0001	0010	1010	1010	1001

Figure 3-1-5   Wall information storage

● Video

Information storage and reading video

Figure 3-1-6   Save data according to orientation

II. Maze information access program control

First start tqd. launch. py in the tqd_Maze robot function package.

```
$ cd ~/Maze robot _ws
$ source install/setup. bash
$ ros2 launch tqd_Maze robot  tqd. launch. py
```

Copy the complete program file contest. py

1. Parameter initialization

```
def valuesinit(self):
    self. mapblock = np. full((16, 16), '00000')
            # 2D array, wall data initialization, flag left
            # down right up
            # flag: whether passed, 0 or 1
            # left: whether there is a road, 0 or 1
            # down: whether there is a road, 0 or 1
```

Chapter3　Maze Robot Simulation Practice | 163

```python
                # right: whether there is a road, 0 or 1
                # up: whether there is a road, 0 or 1

        self.mapblock[0,0] = '10001' # Start wall initialization
        self.crosswaystack = []
        self.mouse_x = 0 # x,y starting point coordinates, these
                         # are the coordinates in the maze
        self.mouse_y = 0
        self.dir = 0             # Maze robot head, 0:up, 1:right,
                                 # 2:down, 3:left
```

2. Coordinate update

```python
    def __coordinateupdate(self):
        if self.dir == 0:            # head up
            self.mouse_y += 1
        elif self.dir == 1:          # head to the right
            self.mouse_x += 1
        elif self.dir == 2:          # head down
            self.mouse_y -= 1
        elif self.dir == 3:          # head to the left
            self.mouse_x -= 1
```

3. Wall data preservation

```python
    def __savewallinfo(self):
        """wall information is saved and modification is
prohibited"""
        frontpath = '1' if self. f_dis > 0.2 else '0'
        rightpath = '1' if self. r_dis > 0.2 else '0'
        backpath = '1'
        leftpath = '1' if self. l_dis > 0.2 else '0'
        # assignment of wall data
        if self.dir == 0:
            self.mapblock[self.mouse_x, self.mouse_y] = ''.
join(('1', leftpath, backpath, rightpath, frontpath))
        elif self.dir == 1:
            self.mapblock[self.mouse_x, self.mouse_y] = ''.
join(('1', backpath, rightpath, frontpath, leftpath))
        elif self.dir == 2:
            self.mapblock[self.mouse_x, self.mouse_y] = ''.
join(('1', rightpath, frontpath, leftpath, backpath))
        elif self.dir == 3:
            self.mapblock[self.mouse_x, self.mouse_y] = ''.
join(('1', frontpath, leftpath, backpath, rightpath))
```

4. Override the parent class turning method

The orientation of the maze robot changes when turning, so override the turnright() , turnleft() and turnback() methods of the parent class.

```python
    def turnright(self):
        self. dir = (self.dir+1)    %4
        return super(Maze_robot, self).turnright()
```

```python
def turnleft(self):
    self.dir = (self.dir+3)    %4
    return super(Maze robot, self).turnleft()

def turnback(self):
    self.dir = (self.dir+2)    %4
    return super(Maze robot, self).turnback()
```

5. Maze robot operation

The ability to move a cell, then update the coordinates, and save wall information.

```python
def moveoneblock(self):
    self.move()
    self.__coordinateupdate()
    if self.mapblock[self.mouse_x, self.mouse_y][0]=='0':
        self.__savewallinfo()
```

After executing the program, the maze robot will move according to the cells. Each time it moves a cell, the coordinates will be updated. If the wall data array flag is '0', the coordinates will be saved.

Task 3   Master the Optimal Path Planning of the Maze

The purpose of this task is to master the principles of optimal path planning and learn how to make contour maps.

I. Principle of making optimal path in maze

Assuming that the maze robot has searched the entire maze or only part of the maze including the starting point and the end point, and has recorded the wall information of each maze cell it has walked, then how does it find a path from the starting point to the end point based on the existing information? What's the optimal path?

The following introduces the concept and production method of contour maps.

Contour map is the abbreviation of contour map. It is just like a general map that can mark the area of the same height, or an isobaric map in a weather report

that can mark the range and size of equal air pressure, as shown in Figure 3-1-7. Then the contour map is used on the maze map to calculate the distance value between each maze grid and the starting point or end point of the maze.

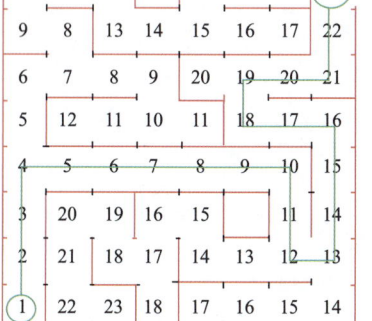

When the maze robot searches and reaches the end point, it has saved all the wall data it has walked through, including the starting point and the end point. Create a contour map with the starting point as the target point. The production steps are as follows:

Figure 3-1-7　Schematic diagram of contour map

① The starting point's contour value is marked as 1, the adjacent and reachable cells' contour value is +1, and so on.

② When the cell has multiple directions to move forward in, calculate the contour values along each direction independently until you encounter a dead end or intersection.

③ If two branches intersect, when calculating the contour value, take the smaller contour value and mark it as the contour value of the intersection cell.

After the contour value is calculated, each cell displays the shortest number of steps from the target point (starting point). The maze robot can return to the starting point as quickly as possible by walking in the direction of decreasing contour value from the end point. At the same time, this is also the shortest path.

The maze robot sprinting from the starting point to the end point is a reverse process that will not be explained again.

II. Maze optimal path program control

First start tqd. launch. py in the tqd_Maze robot function package.

```
$ cd ~/Maze robot _ws
$ source install/setup.bash
$ ros2 launch tqd_Maze robot  tqd.launch.py
```

Copy the complete program file contest. py

1. Make a contour map

```
def mapstepedit(self, dx, dy):
    """create a contour map, modification is prohibited"""
```

Video

Optimal path planning

```python
            self.mapstep = np.full((16,16), 255)
            step = 1
            n = 1
            stack = []
            stack.append((dx, dy))
            cx = dx
            cy = dy

            while n:
                self.mapstep[cx, cy] = step
                step += 1

                count = 0
                # count how many possible directions the current
coordinates can move forward
                if (self.mapblock[cx, cy][-1] == '1') and (self.
mapstep[cx, cy+1]>step):
                    count += 1
                if (self.mapblock[cx, cy][-2] == '1') and (self.
mapstep[cx+1, cy]>step):
                    count += 1
                if (self.mapblock[cx, cy][-3] == '1') and (self.
mapstep[cx, cy-1]>step):
                    count += 1
                if (self.mapblock[cx, cy][-4] == '1') and (self.
mapstep[cx-1, cy]>step):
                    count += 1

                if count == 0:
                    cx, cy = stack.pop()
                    step = self.mapstep[cx, cy]
                    n -= 1
                else:
                    if count>1:
                        stack.append((cx, cy)    )
                        n += 1
                    # randomly pick any direction and it won't affect
                    # the result, because you'll go through them all
                    if (self.mapblock[cx, cy][-1] == '1') and (self.
mapstep[cx, cy+1]>step)    :
                        cy += 1
                        continue
                    if (self.mapblock[cx, cy][-2] == '1') and (self.
mapstep[cx+1, cy]>step)    :
                        cx += 1
                        continue
                    if (self.mapblock[cx, cy][-3] == '1') and (self.
mapstep[cx, cy-1]>step)    :
                        cy -= 1
                        continue
                    if (self.mapblock[cx, cy][-4] == '1') and (self.
mapstep[cx-1, cy]>step)    :
                        cx -= 1
                        continue
```

Chapter3　Maze Robot Simulation Practice | 167

2. Move towards the target point

```python
    def objectgoto(self, dx, dy):
        """move to the target point, modification is prohibited"""
        self.mapstepedit(dx, dy)
        temp_dir = 0
        cx = self.mouse_x
        cy = self.mouse_y
        while (cx!=dx)    or (cy!=dy)    :
            # run along the direction of decreasing contour
            # value until reaching the target point
            step = self.mapstep[cx, cy]
            if (self.mapblock[cx, cy][-1] == '1') and (self.
mapstep[cx, cy+1]<step):
                temp_dir = 0
            elif (self.mapblock[cx, cy][-2] == '1') and (self.
mapstep[cx+1, cy]<step):
                temp_dir = 1
            elif (self.mapblock[cx, cy][-3] == '1') and (self.
mapstep[cx, cy-1]<step):
                temp_dir = 2
            elif (self.mapblock[cx, cy][-4] == '1') and (self.
mapstep[cx-1, cy]<step):
                temp_dir = 3
            d_dir = (temp_dir - self.dir + 4 )%4
            if d_dir == 1 :
                self.turnright()
            elif d_dir == 2 :
                self.turnback()
            elif d_dir == 3 :
                self.turnleft()
            self. moveoneblock()
            cx = self.mouse_x
            cy = self.mouse_y
```

When the maze robot reaches the end or enters a dead end, it will call objectgoto() to return to the previous fork. Walking based on contour maps can greatly reduce running time.

Questions and Summary

1. What are the commonly used search strategies for maze robots?

2. Is there any room for improvement in making optimal paths?

3. The operation of the maze robot involves the conversion between relative and absolute directions. The detected wall information is in relative direction, so it needs to be converted to absolute direction before storage; when reading existing wall information, the absolute direction needs to be converted. Return to the relative direction and control again.

Project 2

The Virtual Simulation Training of the Maze Robot Competition

Learning objectives

1. Understand how to use the virtual simulation evaluation system.

2. Master how to use the virtual simulation evaluation system to conduct a competition.

Task 1　Learn the Virtual Simulation Evaluation System

The purpose of this task is to learn to use the TQD-OC maze robot virtual simulation evaluation system.

I. Function introduction

TQD-OC V2.0 is a new generation of maze robot virtual simulation evaluation system developed and designed by Tianjin Qicheng Science & Technology Co., Ltd. based on the needs of virtual simulation education and teaching, as shown in Figure 3-2-1.

This system deploys functions such as 3D model construction, multi-sensor fusion, intelligent algorithm verification, and real-time data feedback, as

Chapter3　Maze Robot Simulation Practice　169

shown in Figure 3-2-2; it provides first-perspective display and an immersive experience; it supports left-hand algorithms, right-hand algorithm, central algorithm and flood algorithm can easily realize the search of standard virtual maze and the optimization of the shortest path.

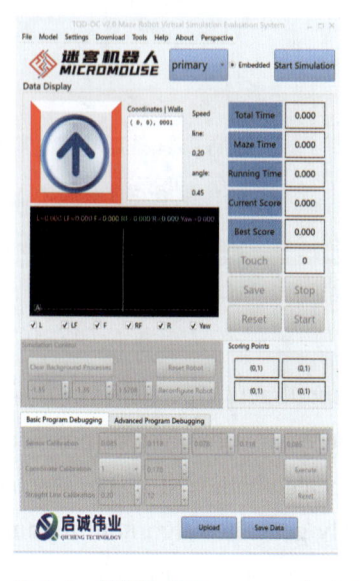

Figure 3-2-1　TQD-OC maze robot virtual simulation competition evaluation system

During the simulation operation, the maze robot's speed, offset, turning angle and other data are fed back in real time to facilitate data analysis by the contestants. This system also provides a performance recording function. The system displays the maze time and running time in real time, records the best results, and automatically sorts the results after the competition. The system supports mode switching. In debugging mode, the control interface will be enabled, and contestants can make the maze robot run more efficiently by correcting parameters. The virtual simulation evaluation system deploys upload and download functions, switching between Chinese and English with one click, meeting the needs of international players and domestic players using the same platform to compete in real time. At the same time, it also provides a large number of 3D maze and maze robot models of different difficulties, which can be freely switched according to competition needs.

| 3D model construction | Multi-sensor fusion | Intelligent algorithm verification | Real-time data feedback |

Figure 3-2-2　Evaluation system features

1. 3D model construction

The evaluation system integrates the function of one-click building of 3D models. It can generate 3D script files from TQD maze model files with one click, making it convenient for users to customize the model, as shown in Figure 3-2-3.

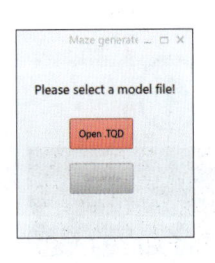

Figure 3-2-3　3D model construction

2. Intelligent algorithm verification

The evaluation system provides the function of intelligent algorithm verification, as shown in Figure 3-2-4. It quickly generates algorithm running trajectories in a flat display mode to assist users in algorithm design.

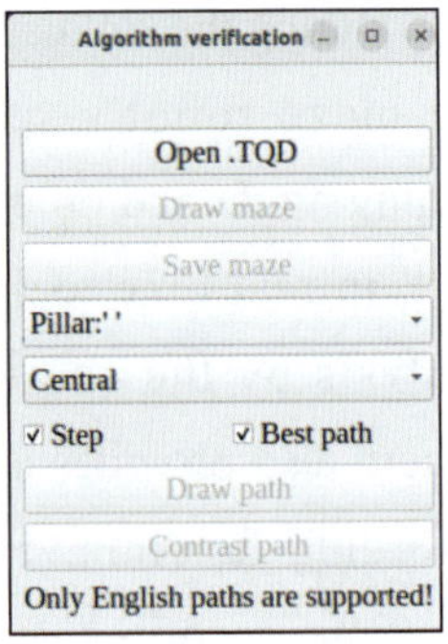

Intelligent algorithm verification tools have the following functions:

① Open. TQD: Load a .TQD maze model file.

② Draw maze: Draw a maze, draw the maze in the model file into a picture.

③ Save maze: Save the picture, save the drawn maze, the running trajectories of various algorithms, etc. to local pictures.

Figure 3-2-4　Intelligent algorithm verification

④ Pillar: The description method of the pillar in the .TQD file, the default is space, which can be changed to '|'.

⑤ Central: Free choice of algorithm, currently supports left-handed algorithm, right-handed algorithm and central algorithm.

⑥ Step: Whether to display contour values, starting point is 1.

⑦ Best path: Whether to draw the best path.

⑧ Draw path: Draw the running trajectory of the robot using the currently selected algorithm.

⑨ Contrast path: Compare the maze original image (Original) , center (Central, green) , right hand (Right, yellow) and left hand (Left, blue) , as shown in Figure 3-2-5.

3. Real-time data feedback

The evaluation system provides real-time data feedback. During the operation of the maze robot, all data are dynamically visualized to facilitate users

Chapter3　Maze Robot Simulation Practice　171

to conduct data analysis, as shown in Figure 3-2-6.

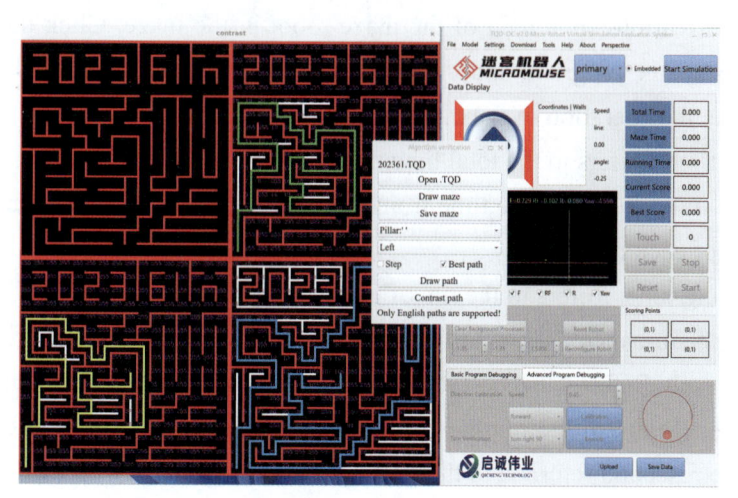

Figure 3-2-5　Comparison of maze trajectories

Number 1: Orientation and wall display.

The arrow indicates the direction of the maze robot's front, and the four frame lines indicate the wall information around the current robot. Red indicates there is a wall, and white indicates there is no wall. The direction of the car's front is one of the important elements in algorithm programming, which helps contestants master the steering changes of the maze

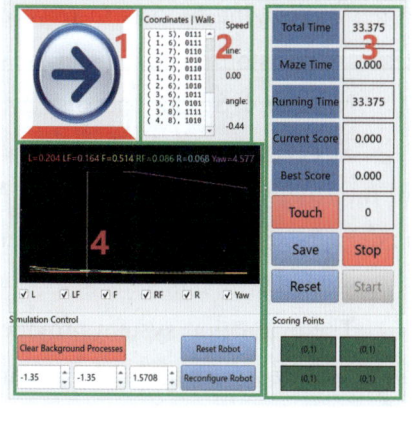

Figure 3-2-6　Data display

robot. The dynamic display of the presence or absence of walls helps contestants understand whether the maze robot detection is correct.

Number 2: Coordinates and wall display.

The coordinate column shows the current position of the maze robot in the maze. In the wall data column, the four numbers correspond to left, bottom, right, and top respectively. There are walls in the four directions. 1 means there is a road in that direction, and 0 means there is no road in that direction. Linear speed and rotation speed are also displayed simultaneously. Wall data can be used as verification as shown in the Figure. The real-time display of coordinates helps contestants verify whether the maze robot has coordinate calculation errors.

Number 3: Scoring system display.

Record various time data of the maze robot in the maze. Click the "Start" button to start the program, and the maze robot will start running; click "Reset" to terminate the operation of the maze robot.

Total time: Refers to the total time the maze robot runs in the maze.

Maze time: Refers to the time from the first startup of the maze robot to the beginning of each run.

Running time: Refers to the run of each run of the maze robot take.

This performance: Refers to the performance achieved during this run.

Best result: Refers to the best result achieved by the maze robot in multiple runs.

Touch: Refers to the number of times the maze robot makes an error and needs to be reset and returned to the starting point to restart.

$$\text{This time's result} = \text{maze time}/30 + \text{running time} + (\text{touch}-1)\times 5$$

In the above formula:Dividing by 30 is to reduce the proportion of search time;Multiplying by 5 is to reward no touches and penalize multiple touches. Users can customize settings in the scoring rules.

Number 4: Sensor data display.

The laser sensor and yaw angle data of the maze robot will be displayed in intuitively drawn waveforms. Assist contestants to debug parameters and understand the deviation of the maze robot.

4. First person display

The evaluation system provides a first-person perspective display function, as if you are in a maze, and you can experience the immersive feeling, as shown in Figure 3-2-7.

II. Operation demonstration

1. System startup

Enter the correct account and password on the login page to jump to the main page, as shown in Figure 3-2-8.

2. Simulation control

Select the appropriate difficulty and click the "Start Simulation" button to open the simulation view, as shown in Figure 3-2-9.

The maze and maze robot currently selected for use can be changed in the model menu.

The evaluation system provides two simulation modes: detached and

● Video

Demonstration of TQD-OC V2.0 evaluation system

Chapter3　Maze Robot Simulation Practice | 173

embedded.

Detached: The simulation view is started independently and takes up more system resources.

Embedded: check the "Embedded startup" radio box, and the simulation view will be embedded in the evaluation system framework and launched in a flat form, effectively saving system resources. It is recommended to choose the embedded startup method.

After the simulation view is opened, you can use the relevant functions in simulation control to control the simulation environment and the maze robot, as shown in Figure 3-2-10.

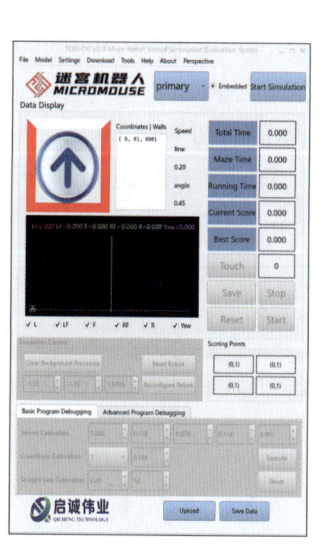

Figure 3-2-7　First person display

Figure 3-2-8　Login page and main page

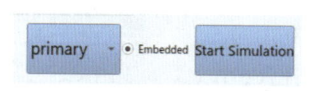

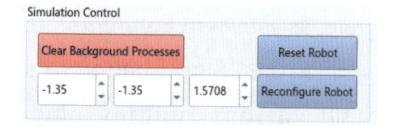

Figure 3-2-9　Open the simulation view

Figure 3-2-10　Simulation control

Clean up background processes: including Gazebo process, debug script process, contest script process, camera script process, etc.

Reset simulation view: Reset the simulation view, including the maze posture and the maze robot posture.

Reset robot position: Re-push the maze robot to appear in the simulation

view at a new position and with a new yaw angle.

3. Program debugging

The evaluation system provides multiple debugging interfaces. After switching the debugging mode in the settings, you can use the relevant functions.

(1) Primary program debugging

Sensor calibration: Setting the relevant thresholds for sensor detection.

Coordinate calibration: Set the size of the unit cell.

For straight-travel calibration: Set the straight-travel speed and the correction coefficient for straight-travel correction of the vehicle attitude, as shown in Figure 3-2-11.

(2) Advanced program debugging

Direction calibration: Calibrate the corresponding relationship between direction and yaw angle.

Turn verification: Control the maze robot to turn and verify whether the turning angle is accurate, as shown in Figure 3-2-12.

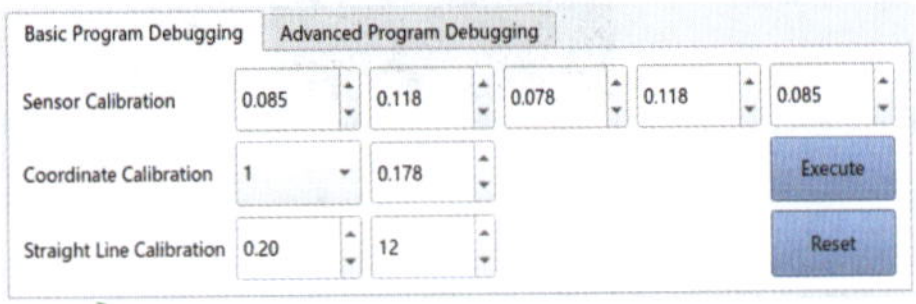

Figure 3-2-11　Primary program debugging

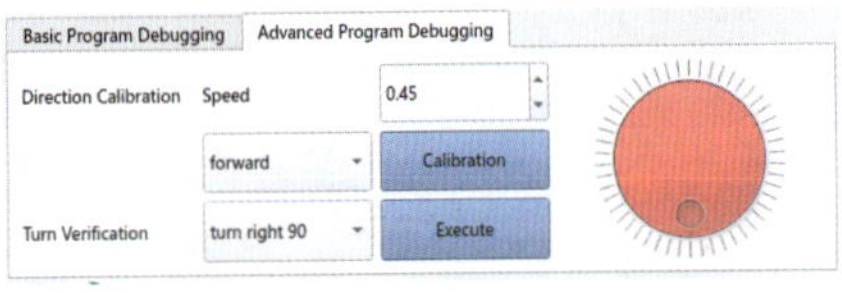

Figure 3-2-12　Advanced program debugging

4. Program running

After debugging the program, switch to competition mode in the settings to run the competition program, as shown in Figure 3-2-13.

Click the "Start" button to start running contest. py, and various running data will be displayed simultaneously in the UI.

① If an error occurs in the maze robot and cannot continue to run, you can click the "Touch" button to terminate the contest. py process.

② Click the "Reset Simulation View" button in the simulation control to restore the maze and robot posture.

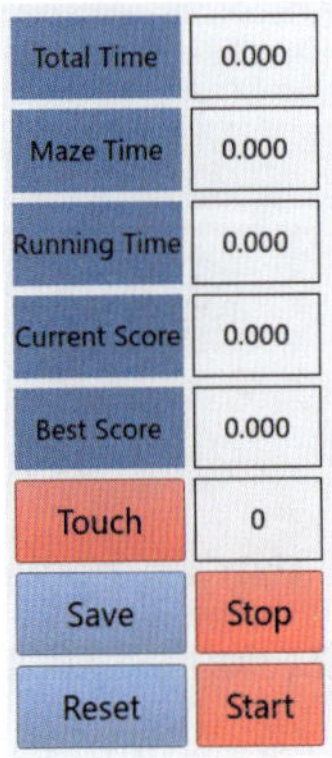

Figure 3-2-13　Scoring system

③ Click the "Start" button again to control the maze robot to run again.

④ Click the "Stop" button to end this competition.

⑤ Click the "Save" button to save this competition data.

⑥ Click the "Reset" button to end this competition and clear all data.

Task 2   Understand the Virtual Simulation Competition

Video

Virtual competition

The purpose of this task is to learn to use the TQD-OC maze robot virtual simulation competition evaluation system to conduct a competition.

I . Model convention and program agreement

The virtual simulation competition is a maze robot competition completed using code. Participants have 100% DIY ability with the code. Therefore, in order to ensure that the competition is fair and smooth, all participants are required to abide by certain rules.

1. Model convention

In the virtual simulation competition, all contestants use the same 3D maze to compete. The maze model is provided by the competition organizing committee, so no special instructions are needed. The maze robot model can use the model provided by the organizing committee, or you can make it yourself, but you must abide by the following conventions:

① The maximum speed of the maze robot shall not exceed 0.4 m/s

② The length and width of the maze robot shall not exceed 0.2 m×0.2 m, and the height shall not exceed 0.05 m.

③ The maze robot is limited to two-wheel differential plug-ins and lidar plug-ins.

④ The initial position of the maze robot in the maze should be set by yourself according to the maze issued by the organizing committee.

2. Program agreement

The competition program file needs to be written in Python and named contest. py. You also need to abide by the following conventions when writing programs.

① Reserve the main program entry, otherwise the program cannot be started.

② Whenever the coordinates of the virtual maze robot change, publish a message on the specified topic:

```
topic name: "transmit",
message type: std_msgs.msg.String
message content: The parameters of the current virtual maze
robot(after changes) are a string separated by commas ',':
"l_dis, f_dis, r_dis, speed_x, speed_z, dir, mouse_x, mouse_y,
mapblock[mouse_x, mouse_y]"
l_dis: horizontal left baffle distance
f_dis: front baffle distance
r_dis: horizontal right baffle distance
speed_x: translation speed
speed_z: spinning speed
dir: heading direction
mouse_x: x coordinate
mouse_y: y coordinate
mapblock[mouse_x, mouse_y]: the surrounding walls use '0' and
'1' to indicate whether there is a road, the length is 4, and the
order is left, lower, right, upper.
```

③ Image processing programs are not allowed. Once discovered, you will be disqualified from the competition.

II. Error troubleshooting

If an error occurs when running the evaluation system or running the competition program, you can refer to the following examples to troubleshoot the error.

1. Error pop-up window

If an error pop-up window appears during operation, please make reasonable modifications according to the pop-up window prompts, as shown in Figure 3-2-14.

Chapter3　Maze Robot Simulation Practice | 177

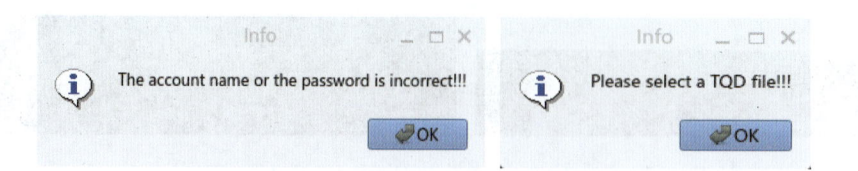

Figure 3-2-14　Error pop-up window

2. Posture offset

If the attitude of the maze robot deviates from the starting position, it is recommended to reset the maze and the maze robot first, otherwise it will affect subsequent operations. Click the "Reset Simulation View" button on the evaluation system panel to reset the maze and maze robot. If it is a separate startup, you can also choose "Reset World" command in the "Edit" menu, or use the shortcut key 【Ctrl+R】 to reset the maze and maze robot, as shown in Figure 3-2-15.

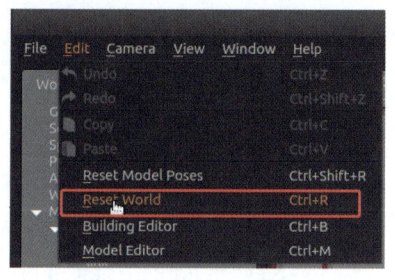

Figure 3-2-15　Reset simulation view

3. The robot has no movement

If you click the "Start" button and the maze robot does not move, please click the "Clear Background Process" button or restart the evaluation system.

4. Parameter is invalid

If the parameters are calibrated but do not take effect in racing mode, it means they are not saved. The calibrated parameters will be saved only after clicking the "Publish Data" button and will take effect when contest. py is run.

5. Program syntax error

When writing Python programs, it is recommended to use VScode or Pycharm, which can prompt most syntax errors, as shown in Figures 3-2-16 to Figure 3-2-19.

Figure 3-2-16　Indentation error

Figure 3-2-17　Extra semicolons

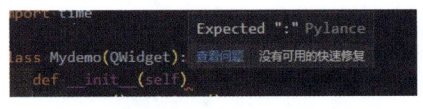

Figure 3-2-18　Missing colon　　　　Figure 3-2-19　Confusion between Chinese and English

6. Program structure error

(1) while statement

In routines, it is mainly used to enter a state, such as straight state or rotation state.

(2) if statement

In the routine, it is mainly used for conditional judgment and what to execute if the conditions are met, such as ending the straight state or rotating state, judging whether there is a wall, etc.

(3) Nested control statement

Nested control statements are used when judging complex situations. At this time, you must pay attention to the hierarchical structure. It is very difficult to troubleshoot such errors.

7. Program parameter error

It needs to be judged based on the actual operating results, such as:

① Unit cell size error: Leads to all coordinate errors, which is very serious.

② Inaccurate sensor threshold: Resulting in inability to walk on the center line, wrong turns, incorrect judgment of whether there is a wall, etc.

③ The turning angle is biased: It will lead to attitude deviation. If the angle is too large or too small, it will affect subsequent operations.

④ Linear speed or rotation speed is too fast: Too fast speed will produce some unpredictable results, affecting subsequent actions, such as hitting the wall directly, spinning in circles, etc.

Questions and Summary

1. How to switch the working mode of the virtual simulation evaluation system?

2. In order for the maze robot to operate stably, its operating parameters must be debugged before the competition.

Appendix

Appendix A

The Rules of the 2022 World Vocational College Skills Competition — the Maze Robot Competition

Ⅰ. Competition form

The competition is held online.

Each participating team sets up a competition venue in their own school or other approved schools, and the competition begins at the designated time. Cameras are set up in the venue to shoot the scene from all angles. Registration and drawing lots are conducted online. One on-site referee and one supervisory arbitrator enter schools to arbitrate and supervise, and one inspector from regulatory authorities will also participate. After the competition, each participating team shall submit the competition results and relevant field monitoring videos within the specified time. Referees give the marks collectively and supervisory arbitrators supervise each team's performance online throughout the entire process.

Ⅱ. Competition content

The competition consists of three modules, including Module A: virtual simulation; Module B: maze racing; Module C: presentation and defense.

Competition session	Module A	Module B	Module C
	Virtual simulation	Maze racing	Presentation and defense
Competition duration	About 3 hours	About 2 hours	About 2 hours
Craftsmanship spirit	At most 10 points will be deducted for lacking professional literacy and safety awareness		

Before the competition, the referee issues a "competition password" online, and then the staff paste relevant information such as the number of lot drawing and competition process in a prominent position on the field. During the competition, modules A, B, and C are proceeded in order. Players use the equipment specified in the competition to operate in the competition field. At the

end of each module, start the next module according to the referee's instructions. When the entire competition is completed, the contestants, under the supervision of the on-site referee, upload the video files to the designated competition platform within the specified time for the referee to evaluate.

The correct "competition password" must be displayed throughout the entire process in cameras and video recorders.

III. Competition process

According to the time set by the competition organizing committee, the registration starts at 8:00 and the competition starts at 8:30 and last for 420 minutes.

The process includes registration, virtual simulation (about 3 hours), maze racing (about 2 hours), presentation and defense (about 2 hours), and scoring. The specific arrangements are as follows.

Date	Time	Work items	Participants	Manner
The day before the competition	10:00-12:00	Referee conference	Chief referee, referees, supervisory arbitrator	Online and offline
	13:00-14:00	Team leader meeting	Team leaders, chief referee	Online
	14:00-16:00	Competition environment test	Team leaders, on-site referee, supervisory arbitrator	Online and offline
	16:00	Closed competition environment test	Chief referee, supervisory arbitrator	Online
Competition day	8:00-8:30	Contestant registration and draw lots	Contestants, encrypted referee, support team	Online and offline
	8:30-11:30	Module A virtual simulation	Contestants, on-site referee, chief	Online
	11:30-13:00	Lunch		
	13:00-15:00	Module B maze racing	Contestants, on-site referee, chief referee, support team	Online and offline
	15:00-17:00	Module C presentation and defense	Contestants, on-site referee, chief referee, support team	Online
	17:00-19:00	Scoring and score check	Referee, supervisory arbitrator	Online and offline

IV. Score evaluation

1. The chief requirement table

Referees and supervisory personnel work both collectively and separately. There are one chief referee, one arbitrator, six scoring referees, 11 on-site referees, 11 on-site supervisory arbitrators and 11 on-site disciplinary inspectors .

Appendix | 181

The chief referee, supervisory arbitrators, and scoring referees work offline collectively. There is one on-site referee, one supervisory arbitrator and one inspector from regulatory authorities at the competition site in each of the 11 participating schools.

On-site referee and on-site supervisory arbitrator: Selected in the province/ city by the World School Competition executive committee (those who are directly related to the participating school should be avoided) , responsible for the entire process of arbitration and supervision at the competition site.

Scoring referees: Selected from the competition referee pool by the World School Competition executive committee, with a proportion of provincial and municipal referees being no less than 50%.

On-site disciplinary inspectors: dispatched by the provincial department where the participating school is located.

After the competition, the scoring referees will score back-to-back (including automatic scoring by the machine evaluation system) , and in principle, the experts and scoring referees should work collectively in the host school.

No.	Referee type	Required number	Working venue
1	Chief referee	1	The host school
2	Arbitrator	1	The host school
3	Scoring referee	6	The host school
4	On-site referee	11	Participating schools
5	On-site supervisory arbitrator	11	Participating schools
6	On-site disciplinary inspector	11	Participating schools

2. Evaluation method

Both offline and online evaluation are used to score modules A, B, and C.

(1) Scoring starts immediately after the competition. During the scoring process, all participating teams must leave and no further operations are allowed. Technical support personnel are waiting for the scoring referees' instructions on the field. The competition equipment remains the way in which all contestants leave the site.

(2) The scoring method of Module A : Scoring referees and on-site referees cooperate to monitor the on-site camera and observe the contestant's operation process. They give marks according to the virtual simulation process and its results. Supervisory arbitrators supervise the entire process.

(3) The scoring method of Module B: Contestants participate in on-site maze racing according to the referee's instructions. Technical support personnel adjust monitors according to referees' requirements to ensure that there are no blind spots. Supervisory arbitrators supervise the entire process.

(4) The scoring method of Module C: Contestants present and defend their work according to referees' instructions, and marks are given according to the scoring criteria.

V. Competition arrangement

1. Preparation before the competition

7 days before the competition, each participating school confirms the competition venue, reports it to the executive committee of World Vocational College Competition , and completes preparation work such as personnel allocation, equipment debugging, and environmental layout.

Participants: participating schools, support teams, liaison officers, and team leaders.

2. Event briefing

Hold an event briefing 15 days before the competition to announce the competition time, competition method, environmental requirements, competition process and announcements.

Participants: experts, chief referee, team leader, and guidance teachers.

3. Confirmation of the competition venue

One day before the competition, the competition environment of the participating school will be tested. The experts, referees, and supervisory arbitrators of the Competition executive committee conduct inspection, acceptance, and testing through a Tencent meeting (the meeting number will be issued by the participating school liaison and confirmed by the support team through SMS). Once approved, the venue will be sealed and a relevant video will be recorded.

Participants: participating school leaders and liaison officers, experts, referees, supervisory arbitrators, and support teams.

4. Preparation for entry

Before the designated time on the competition day, all participating schools and relevant personnel shall enter the competition venue. The support team shall create a video conference for the competing team, and notify the contact person of participating schools in the same team of the video conference number via SMS. The contact person replied with the designated mobile phone number:

"× × × (school name) has received the Tencent conference number for the maze robot event: × × × × × × × × × ×. All preparations must be made before × o'clock in the morning/afternoon of × month × ×. We hereby confirm. " The on-site referees open the venue and record a video. Participating schools enter the corresponding competing team through a video conference and debug all equipment. Only contestants, on-site referees, on-site supervisors, technical support engineers of cooperative enterprises, technical personnel for video shooting, conversion, uploading, and video connection shall be present at the competition venue. Throughout the competition, referees must maintain video connectivity and be able to monitor the competition venue. Participating teams log in to the video conference on time and use the match number as their conference name. The external wide-angle camera are turned on until all competition-related matters are completed. The support team, on-site referees, and supervisory arbitrators inspect the competition venue and on-site personnel.

5. Identity verification

At 8:00 am on the competition day, each participating team shall be connected to the support team through a video conference within the specified time. Contestants shall make their appearance, show their ID cards (passports) , student ID cards, and instructor ID cards in front of the camera. The support team take screenshots to complete the identity verification.

6. Determine the match number

Encrypted referees draw lots to determine the match number of each participating team under the supervision of the supervisory arbitrators; Each participating team uses their match number to enter the specific Tencent meeting. Team leaders of participating schools reply with a text message to confirm.

7. Real time recording

The support team publish the "specific competition logo" at the same time, which will be fixed and posted (or written) by each participating school in a visible position during video recording.

8. Upload the competition results

At the end of the competition, contestants upload the competition results to the designated address within the specified time according to the requirements.

9. Upload the competition video

Name the recorded video files as "W12+maze robot event +time (close and open the venue) +match number+workstation number. MP4". The competition video clip should be in MP4 format, and any editing or dubbing are forbidden. There

is no specific requirement for the video recording software and the video shall be compressed using H. 264/AVC (MPEG-4 Part 10) encoding format. The bitrate of the dynamic bitstream shall not be less than 1 024 kbit/s; resolution set to 720 × 576 (standard definition 4 ∶ 3) or 1 280 × 720 (high-definition 16 ∶ 9) ; progressive scanning (frame rate of 25 frames/second). The audio is compressed using AAC (MPEG4 Part3) format; sampling rate 48 kHz; bitstream 128 kbit/s (constant). Video clips must be uploaded to the designated email address on time (taken charge by the host school). The deadline for the on-site competition video uploading is within 60 minutes after the competition. The video clips of opening and closing competition venues shall be uploaded before 12:00 pm on the competition day.

Module	Category	Naming format
Module A	recording	W12+Maze robot event +Module A recording+the workstation number. MP4
	The recording of the back of workstation	W12+Maze robot event +the recording of the back of Module A workstation+the workstation number. MP4
	The recording of the whole workstation	W12+Maze robot event +the panoramic recording of Module A+the workstation number. MP4
Module B	Mobile recording	W12+Maze robot event +the mobile recording of Module B+match number. MP4
	Panoramic recording	W12+Maze robot event +the panoramic recording of Module B+match number. MP4
	The recording of the scoring system	W12+Maze robot event +the recording of Module B scoring system+match number. MP4

10. Completion of the competition

After the competition and the uploading of accurate video clips, team leaders reply to the designated mobile phone number to confirm: "The maze robot competition ××× (match number) has completed the competition, and hereby confirms. "

11. Scoring

According to the competition requirements, scoring referees evaluate the works (including machine evaluation - automatic scoring). The evaluation results are compiled and summarized by the chief referee.

12. Score calculation and publication

The scoring method shall be in accordance with the competition regulations, and the final score shall be reviewed by the supervisory arbitrators. The results will be publicly announced according to the requirements of the organizing committee.

Appendix B

International Training Course Standards for "Simulation and Design of Maze Robot"

Module name: Simulation and Design of Maze Robot.

Total classes: 48; Theoretical classes: 16; Practical classes: 32.

Course category: professional elective courses.

Pre-requisite course: unlimited.

Targeted majors: electronics and information, equipment manufacturing.

I. Course orientation and objectives

1. Course orientation

In accordance with the innovation practice curriculum reform in colleges and universities, higher vocational colleges and technical secondary schools, the course emphasizes independent creativity, hands-on practice, electromechanical control, sensing and other technologies. The course boasts progressiveness, enlightening and practicality.

Using the maze robot as a carrier, the course carry out practical teaching activities, aiming to strengthen students' hands-on abilities, enable them to proficiently master various mechanisms and complete the predetermined robot programming. The course also guides students to develop and create reasonable mechanisms and to propose solutions to complex engineering problems. Besides, the course also familiarizes students with the design regulations of electromechanical systems, assists them in designing required electromechanical products and systems and motivates them to demonstrate innovative awareness in mechanical structure design, circuit design and software programming.

2. Course objectives

The course enables students to understand the composition and operation of mechatronics equipment through experiments. Students need to analyze and design the structure of mechatronics systems, and have a certain understanding in techniques so as to improve hands-on and innovative skills.

(1) Knowledge objectives

① Understand the basic mechanical structure and characteristics of robots;

Understand the basic concepts, classifications, and characteristics of electrical and pneumatic control industrial technology; Understand the working principle, types, and practical application methods of electrical and pneumatic components, as well as the composition and structure of electrical and pneumatic control systems; Understand the application of basic electrical and pneumatic control technologies; Understand the structure and characteristics of modern automation control systems.

② Master key technologies, core knowledge, and core skills such as embedded microcontrollers, sensors, intelligent control algorithms, 3D virtual simulation, and artificial intelligence.

(2) Capability objectives

① Master robot virtual simulation technology, systematically simulating robot modeling (URDF) , creating simulation environments (Gazebo) , and perception environments (Rviz)

② Master the python programming language, learn and practice.

③ Implement the construction, communication, and related algorithms on a virtual platform based on the maze robot virtual simulation system.

(3) Objectives of ideological and political education

① Elaborate on how to discover scientific problems from the perspective of dialectical materialism. In this way, students receive preliminary training in logical thinking, critical thinking, and master the ability to analyze and solve problems. Cultivate students' dialectical materialism worldview and motivate them to solve problems in a scientific way.

② Elaborate on the contributions and academic demeanor of Chinese scientists in the field of robot control, enhance students' cultural confidence and national pride, publicize the spirit of craftsmanship and core values of socialism, and guide students to become firm believers, active disseminators, and practitioners of those core values.

③ Elaborate on the shortage of talents in the field of robot control, as well as the urgent demand for advanced technology in China, motivate students to strive for national development, stimulate and enhance students' national self-esteem and sense of responsibility.

Ⅱ. Approaches and methods for curriculum objectives

No.	Content	Overview	Credit hour	Teaching method
1	The origin and development of maze robot	Understand the origin, use and types of maze robot	2	Instruction, demonstration and discussion

Continued

No.	Content	Overview	Credit hour	Teaching method
2	Basic knowledge and control principles	Understand the structure, function, and purpose of basic units; understand the working principles and construction process of commonly used structures; understand the types, characteristics, functions, and uses of commonly used controls; understand the control principles and processes of several commonly used structures	2	Instruction, demonstration and discussion
3	The virtual simulation system, robot ROS system, Python programming	Learn maze modeling, robot posture control, maze search algorithms	18	Instruction, demonstration, practice and discussion
4	Designing creative models	Design and operate maze robots. During the design process, students need to be diligent, meticulous, rigorous, patient, and persistent in all aspects. It is aimed at cultivating students'ability to analyze and solve problems and the ability to work in a team, as well as cultivating their spirit of exploration, innovation, and pursuing truth	24	Practice
5	Summary	Check and summarize the completed model and its function, demonstrate the innovation points and experience by PPTs. Cultivate and strength communication and expression skills	2	Check and summarizing

Ⅲ. Assessment methods and performance evaluation

1. Assessment type

Examination.

2. Assessment methods

Evaluation content	Evaluation methods	Evaluation criteria	Proportion in total score
Formative assessment	Attendance, classroom discussion and assignments	The role students play in the project, their work attitude and attendance	40%
Final examination	Products + presentation	Idea innovation, structure feasibility, effect visibility, scientific and standard experimental reports	60%

Ⅳ. Recommended textbooks and references

1. Self compiled handouts

2. References

(1) Wang Chao, Gao Yi, and Song Lihong. *Micromouse Design Principle and Production Process*. China Railway Press, 2019.

(2) Yang Chenguang, Li Zhijun, and Xu Yang. *Robot Simulation and Programming*. Tsinghua University Press, 2018.

(3) Li Yansheng, Yang Meimei, Wei Bo, et al. *Modeling and Simulation of Robot Systems*. Beijing University of Posts and Telecommunications Press, 2020.